Oxford Applied Mathematics
and Computing Science Series

General Editors
J. Crank, H. G. Martin, D. M. Melluish

W. W. SAWYER
Professor Emeritus, University of Toronto

A first look at Numerical functional analysis

1978

CLARENDON PRESS · OXFORD

Oxford University Press, Walton Street, Oxford OX2 6DP

OXFORD LONDON GLASGOW NEW YORK
TORONTO MELBOURNE WELLINGTON CAPE TOWN
IBADAN NAIROBI DAR ES SALAAM LUSAKA ADDIS ABABA
KUALA LUMPUR SINGAPORE JAKARTA HONG KONG TOKYO
DELHI BOMBAY CALCUTTA MADRAS KARACHI

British Library Cataloguing in Publication Data

Sawyer, Walter Warwick
    A first look at numerical functional analysis.
    —(Oxford applied mathematics and computing science series).
    1. Numerical analysis   2. Functional analysis
    I. Title   II. Series
511'.7'015157   QA297   77-30689

ISBN 0-19-859628-6
ISBN 0-19-859629-4 Pbk

Typeset in Northern Ireland
at The Universities Press, Belfast
Printed in Great Britain
by Richard Clay & Co. Ltd., Bungay

Preface

THE AIM of this book is to provide an intelligible introduction to functional analysis by giving samples of its applications to numerical analysis. I have been very much helped in my attempt to select topics of practical value by discussions and correspondence with mathematicians in the University of Lancaster, Brunel University, the University of Bristol, the University of Glasgow, the University of Strathclyde, and the University of Exeter. For this help I should like to express my sincere thanks.

34 *Pretoria Road,*
Cambridge.
July 1977

Contents

Notes on symbols and terminology

FOR CONCEPTS not mentioned here, see the general index.

$\rightarrow$,   (i) 'tends to the limit', (ii) 'maps to', e.g. 'input $\rightarrow$ output'.

$\equiv$,   'identically equal'. An *identity*, $f(x) \equiv g(x)$, indicates that the functions have the same values for all relevant x.

$\simeq$,   'approximately equal'.

$\sim$,   see section 8.4.

!,   factorial; $n! = n(n-1)(n-2) \ldots 3 \times 2 \times 1$.

Δ,   finite difference operator, 'the change in'.

D,   differentiation operator, d/dx.

ln,   natural logarithm, $\log_e$.

$P_n(x)$,   Legendre polynomial, section 8.4.

$T_n(x)$,   Chebyshev polynomial, section 8.4.

$[a, b]$,   the closed interval, corresponding to $a \leqslant x \leqslant b$, in contrast to the open interval, (a, b), corresponding to $a < x < b$.

$\mathbb{R}$,   the real numbers

$\mathbb{C}$,   the complex numbers.

$\in$,   'belongs to'. Thus, $x \in \mathbb{R}$ means 'x is a real number'. Note that $\in$ is not the same symbol as the Greek epsilon, ε, traditional symbol for a small, positive number.

$\{:\}$,   a collection of objects satisfying a condition. Thus $\{x : x > 0\}$ means 'the positive numbers' and $\{(x, y) : x^2 + y^2 = 1\}$ means 'the collection of points for which $x^2 + y^2 = 1$', that is, the unit circle.

$\max\{:\}$,   the maximum, the largest number in the collection indicated.

$\min\{:\}$,   the minimum, the smallest number in the collection.

$\sup\{:\}$,   the supremum, the 'ceiling' of the collection, that is, the smallest number not exceeded by any other in the collection. Thus $\sup\{1/2, 2/3, 3/4, \ldots, n/(n+1), \ldots\}$ is 1. We cannot use 'max' here, since there is no largest number in this collection.

$\inf\{:\}$,   infimum, the 'floor' of the collection.

majorant, something known to be larger than the object under investigation.

$\Rightarrow$, 'implies', e.g. '$x = -5 \Rightarrow x^2 = 25$' means 'if $x = -5$, then $x^2 = 25$'.

$\Leftrightarrow$, connects statements, each of which follows from the other, e.g. $2x = 10 \Leftrightarrow x = 5$.

$\ell_1, \ell_2, \ell_\infty$, see section 2.3.

ℓ_p, section 4.1.

$\mathcal{L}_2$, section 8.3.

$\mathscr{C}[a, b]$, section 3.3.

$\mathscr{X}, \mathscr{Y}, \mathscr{Z}$, symbols for spaces.

$\mathscr{B}(\mathscr{X}, \mathscr{Y})$, the space of bounded, linear operators, $\mathscr{X} \to \mathscr{Y}$, section 5.3.

$\mathscr{X}^*$, the space dual to $\mathscr{X}$, section 10.1.

$\mathbb{R}^n$, vector space of n dimensions with real co-ordinates. If $u = (u_1, \ldots, u_n)$ and $v = (v_1, \ldots, v_n)$ are in $\mathbb{R}^n$, then $u + v = (u_1 + v_1, \ldots, u_n + v_n)$ and, for any number, k, $ku = (ku_1, \ldots, ku_n)$.

$\mathscr{E}^n$, Euclidean space of n dimensions.

$u \cdot v$, scalar product in $\mathscr{E}^3$, sections 8.1 and 9.1.

(u, v), generalization of $u \cdot v$, chapter 8.

Perpendicular projection. If $PM \perp OMQ$, the vector OM is called the perpendicular projection of the vector OP onto the line OQ.

$| \ |$, absolute value; the magnitude of a number, irrespective of its sign, e.g. $|-7| = |7| = 7$.

$\|PQ\|$, the length of the line segment, PQ. Its generalizations; vector norm, $\|v\|$, section 3.2; $\|v\|_p$, norm of a vector in ℓ_p; operator norm, $\|A\|$, section 5.2; function norm, $\|f\|$, section 3.3; norm in $\mathcal{L}_2$, (Hilbert space), section 8.3; $\|A\|_p$, norm of an operator A, $\ell_p \to \ell_p$ or $\ell_p \to \mathbb{R}$.

$d(P, Q)$, the distance of P from Q.

$S(P, r)$, the sphere, $B(P, r)$ the open ball.

$\bar{B}(P, r)$, the closed ball, in each case with centre P and radius r. See section 2.3.

I, the identity operator or matrix, 'leave everything as it is'.

A^{-1}, the operator or matrix inverse to A. When A^{-1} exists, $A^{-1}A = AA^{-1} = I$. See section 5.6.

Invariant line. A line such that every point on it maps, under a specified linear transformation, M, to a point of the same line.

Eigenvector. A non-zero vector, v, lying in an invariant line; it satisfies the equation, $Mv = \lambda v$, for some number λ.

Eigenvalue. The number, λ, mentioned in the definition of eigenvector.

1 A first course in functional analysis

IN THE introduction to his book, *Functional analysis and numerical analysis*, L. Collatz said that numerical analysis had been revolutionized by two things—the electronic computer and the use of functional analysis.

This statement is striking not only for its emphatic tone but also for the contrast of mathematical epochs it involves. Numerical analysis is an earthy subject concerned with questions involving numbers. Even if it uses very sophisticated modern equipment, essentially it is concerned with arithmetic, the oldest branch of mathematics, whose beginnings lie in prehistory. Functional analysis, on the other hand, while it has roots in nineteenth century mathematics, is a product of the present century.

In trying to cope with such a modern branch of mathematics, a student of numerical analysis encounters two difficulties. Any part of modern mathematics is the end-product of a long history. It has drawn on many branches of earlier mathematics, it has extracted various essences from them and has been reformulated again and again in increasingly general and abstract forms. Thus a student may not be able to see what it is all about, in much the same way that a caveman confronted with a vitamin pill would not easily recognize it as food.

The second difficulty is that functional analysis was not created with numerical applications in view. It arose from a great variety of sources—from the calculus of variations, from integral equations, from Fourier series, from mechanics, from the theory of real and complex variables, from number theory, and from other topics. It therefore has a great range of possible applications and a student cannot assume, because a theorem in functional analysis is generally regarded by mathematicians as of great importance, that it will necessarily help us in problems of numerical analysis.

The aim of this book is to make some contribution towards overcoming these two difficulties—by explaining the ideas of the subject and, as far as possible, by emphasising those ideas that have proved useful to numerical analysts. In the main the concepts of functional analysis will be introduced by discussing problems in numerical analysis from which these concepts could arise. References will be given so that readers can go back to original sources to clear up any obscurity or to judge for themselves the relevance of the topic to their own interests. A reference such as (Smith, (3), p. 42) would indicate page 42 of the third paper by Smith in the list of references at the back of the book.

Also at the front of the book will be found a list of symbols and words used, with their meanings.

This list should be consulted if any unfamiliar word or symbol is encountered. It seemed better to arrange things this way, rather than to risk wearying readers by explaining in the body of the book matters which were already familiar to them. Indeed at present it is extremely difficult to know what is familiar. This is partly owing to the separation of school programs into modern and traditional, and, at a deeper level, to the fact that mathematics is becoming so varied that it is no longer possible to assume that some agreed core is known to all students of mathematics.

In many schools, a modern course, such as S.M.P., is supplemented by more traditional material, designed to give facility in algebraic manipulation. This is probably the ideal background for our subject. This book does assume some acquaintance with vectors and matrices. Students with a traditional background (which they will find an asset in this work) may find it helpful to consult the S.M.P. 'transitional' books, published at a time when schools were just beginning to move from the older to the newer syllabus. The introduction to matrices in S.M.P. Book T.4 is perhaps better than anything in the later S.M.P. publications.

It is not too difficult for someone who has a good understanding and command of traditional algebra and calculus to acquire the more modern concepts. This is perhaps because the later developments grew on the foundation of the earlier ones. Students who are in the opposite position, of having the more recent concepts, but not feeling comfortable with the older topics and skills, will probably find it harder to adjust and establish a correct balance of new and old. Students, who find difficulty with the manipulative aspects of the work, should consult, and above all work exercises from, books of an earlier period when such skills were emphasized—often to excess.

1.1. 'Soft and hard' analysis

There is a custom among mathematicians of referring to classical analysis as 'hard' and modern analysis as 'soft'. Students may not be surprised to hear nineteenth-century analysis described as hard, but one would expect the twentieth-century analysis that followed it to be more advanced and even harder. The explanation of this terminology is along the following lines. Classical analysis frequently involved long chains of reasoning and calculation. Mathematicians tried to simplify it by disentangling the various strands involved in it, in order to find simplicity underlying the complexity. The real numbers have many properties;—algebraic properties, properties concerned with limits, order properties such as that 3 comes before 10 when we count. An effort was made to separate these and to see what could be said about a mathematical structure if you knew only its algebraic properties, or its order properties, or its behaviour in regard to limits. Such studies, concerned only with one aspect of the real numbers, were naturally simpler than the older work that dealt with all aspects simultaneously. Sometimes they did not involve calculation at all. As P. J. Davis said, functional analysis 'loves soft analysis and avoids hard analysis like the plague; its ideal proof is wholly verbal' (see Hayes, p. 160).

The advantage of soft analysis, besides its relative simplicity, is its generality. One of the main features of mathematics in the last century has been the gradual realization that the theorems and procedures of algebra and calculus apply not only to the real and complex numbers, but to a wide variety of other objects, including several that are of great interest to numerical analysts. All the work of this book will illustrate this theme.

Ideas of great generality are extremely valuable, but they are hardly ever sufficient, by themselves, for dealing with a particular, concrete situation. Soft analysis therefore is a supplement to, not a substitute for, hard analysis. Soft analysis, as we have just seen, grew from classical analysis and revealed wider applications of the older ideas. It therefore binds together our knowledge of different topics and reduces the strain on the memory. Instead of learning disconnected facts about unrelated objects, we can take a classical theorem or procedure and see it operating again and again in different environments. The organization of this book corresponds to that idea. We shall take in turn concepts or theorems of classical mathematics, briefly review their role in their original setting (real or complex numbers) and then see to what other situations they can be applied.

Functional analysis helps us also by providing a way of visualizing what we are doing. The whole language of the subject is in terms of 'spaces'. This means that we are able to use geometrical and pictorial imagery in situations that, at first sight, appear to have nothing to do with geometry.

What has been discussed so far may be called the inherent benefits of functional analysis. Functional analysis has also acquired a social value—it has become necessary for communication. Books on numerical analysis, even when using classical methods, often express these in modern terminology. For instance Cheney, on page 85 of his well known book, *Introduction to approximation theory*, discusses the problem of finding a polynomial that closely approximates a given continuous function. The argument is purely classical, involving inequalities for real numbers. The terminology however involves two items that are not classical; functions are defined on 'a compact metric space' and the algebra repeatedly uses $\|P\|$, the symbol for a norm. To follow the argument on this page, a student needs only to know the meaning of these concepts and their most elementary properties.

2 Old ideas in new contexts

2.1. Examples of iteration

As HAS already been indicated, we plan to extend several concepts of classical analysis so that they become available for a wider range of applications. A frequent object of such applications is the process of iteration, one of the most valuable and widely used methods of calculation. It is therefore helpful to look at a few examples of iteration, in order to see something of the variety that this process covers.

Example 1. The idea of iteration is extremely old and is implicit in Zeno's paradox of Achilles and the tortoise (B.C. 500). The tortoise is given unit distance start but Achilles moves ten times as fast. When Achilles has covered a distance x, the tortoise is at the point t, where $t = 1 + (0.1)x$. Initially $x = 0$, $t = 1$. When Achilles has reached the tortoise's initial position, $x = 1$, but by then $t = 1.1$. When x is raised to 1.1, t has become 1.11 and so the argument continues. Its effect is to produce a sequence of x-values, say $x_1, x_2, x_3, \ldots$ with $x_{n+1} = 1 + (0.1)x_n$. As $n \to \infty$, x_n approaches the solution of $x = 1 + (0.1)x$.

Example 2. In any application of iteration convergence has to be considered. We would of course get disastrous results if we tried to solve $x = 1 + 3x$ by the iteration $x_{n+1} = 1 + 3x_n$, which would lead to $+\infty$ rather than -0.5.

Example 3. It is not necessary that the equation should be linear. If we take $x_{n+1} = 3/(x_n + 10)$ with $x_0 = 0$, six iterations are sufficient to give 0.291 502 622 as a solution of $x^2 + 10x - 3 = 0$.

Example 4. If we perform the iteration

$$x_{n+1} = x_n y_n + x_n + 0.07$$
$$y_{n+1} = x_n^2 + y_n^2 + y_n - 0.41$$

with initial values $x_0 = 0$, $y_0 = 0$, about twenty iterations are sufficient to give $x = 0.111\,002\,285$, $y = -0.630\,617\,546$ as an intersection of the curves $xy = -0.07$, $x^2 + y^2 = 0.41$.

Needless to say, this is not offered as an example of practical value. The examples in this book have been kept within the capacity of a pocket-size programmable computer. In a serious, real-life application there might well be a hundred equations in a hundred unknowns.

Example 5. The function $f(x) = e^{-x}$ satisfies the integral equation

$$f(x) = 1 - \int_0^x f(t)\, dt.$$

It can be obtained by the iteration

$$f_{n+1}(x) = 1 - \int_0^x f_n(t)\, dt.$$

If $f_0(x) = 0$, $f_1(x) = 1$, $f_2(x) = 1 - x$, $f_3(x) = 1 - x + (x^2/2)$ and so on. The terms of the series for e^{-x} gradually appear.

Here again we have taken a very simple example from a very wide class. A classical method for dealing with integral equations of the form

$$f(x) = g(x) + \int_a^b K(x, y) f(y)\, dy,$$

where $g(x)$ and $K(x, y)$ are given, is to apply iteration to determine $f(x)$. The solution then appears as a series known as the Neumann series.

Example 6. The integral equation in example 5 is a linear integral equation. It is not necessary to restrict ourselves to this type. If we have the equation

$$f(x) = x + \int_0^x [f(t)]^2\, dt$$

and use the iteration

$$f_{n+1}(x) = x + \int_0^x [f(t)]^2\, dt$$

with $f_0(x) = 0$ initially, we find $f_1(x) = x$,

$$f_2(x) = x + (x^3/3), \quad f_3(x) = x + (x^3/3) + (2/15)x^5 + (1/63)x^7.$$

In fact $f(x) = \tan x$ is the solution of our equation, and in the nth iterate, $f_n(x)$, the first n terms coincide with the Taylor series for $\tan x$.

These six examples differ in the material they involve. In the first three examples we are concerned with a single number, x; example 4 deals with a pair of numbers (x, y) which may be considered as representing a point in a plane; examples 5 and 6 are concerned with functions of a single variable. But it is clear these examples have something in common, and this is particularly evident if we imagine the iteration being carried out by a computer. In each case there is a sub-routine which is applied again and again, the output of each stage becoming the input of the next. All our examples can be expressed by a single symbolic form. If T denotes the sub-routine and ϕ_0 the initial input, we have successive outputs ϕ_1, ϕ_2, $\phi_3 \ldots$ with $\phi_1 = T\phi_0$, $\phi_2 = T\phi_1$, and so on. We may write $\phi_n = T^n\phi_0$ to indicate that the nth output is obtained by applying n times the process T to the initial input ϕ_0.

When we study iteration from the viewpoint of functional analysis, we are not concerned about the nature of the objects ϕ_n. They may be numbers, vectors, matrices, functions, or maybe something else altogether. We are trying to discover properties that relate to the process of iteration and that can be used, in a wide variety of circumstances, to distinguish cases, like example 2 above, in which iteration leads to a disaster, from those in which it proves reliable. You may well doubt whether such a theory is possible and indeed if we took it in its most general form—a study of all repeatable operations—there would indeed be little to say. What has happened historically is that particular examples of successful iteration were found before any general theory was envisaged. From this experience of successful iterations mathematicians managed to winnow certain general theorems and principles to guide them in their future work. It is with such results that we shall be concerned.

2.2. Functions, ancient and modern

The present use of the word *function* is in most respects much wider than it was in former times, though in one respect it is narrower. Three centuries ago an equation $y = f(x)$ would have implied that x and y were numbers, and that the calculation giving y in terms of x belonged to a class that was conventionally accepted.

In nineteenth-century work on complex variables, a large, and very beautiful, part was played by *many-valued functions*, for example $y = f(x)$ being defined by $y^2 = x$. At the present time the first type of restriction has disappeared entirely. It is no longer required that x and y in $y = f(x)$ be numbers, nor is there any restriction on the kind of calculation or rule that leads from x to y. But the term *many-valued function* has been discarded; the one stipulation that is made is that to an acceptable input, x, there is one and only one output, y.

In all the six examples of iteration, the symbol T thus represents a function, for in each of these examples the input determines without any uncertainty the output.

At some stage of life this must involve a student in some mental readjustment. A student in the sixth form, who has been asked to give an example of a function, might mention $f(x) = \sin x$ or $f(x) = x^2$. It would probably cause some surprise if the student mentioned the operations of integration and differentiation. Yet, with modern usage, such a suggestion might well be justified. Suppose, in order to avoid analytical complications, we agree that the acceptable inputs are to be polynomials. To any input, $P(x)$, the operation of differentiation makes correspond one clearly defined output, $P'(x)$. Equally, if by integration we understand the calculation of some definite integral, this operation too defines a function; to any polynomial input, $P(x)$, there corresponds without any uncertainty or ambiguity the output $\int_0^1 P(x)\, dx$.

Functions are often indicated by the symbolism of an arrow. Thus instead of $f(x) = x^2$ we may write $f : x \to x^2$. This indicates that when the input is any number x, the output is the square of that number, x^2. This type of symbolism is also used for another purpose, to indicate the kind of things that occur as inputs and outputs. Thus, if our example of squaring is concerned with real numbers, we may write $f : \mathbb{R} \to \mathbb{R}$, where $\mathbb{R}$ is a rather pretty symbol used to indicate the real number system. This indicates that the input and output are real numbers.

In the same way, for the function D that represents differentiation, we may define D by writing $D : P(x) \to P'(x)$. We may also indicate that the input and output are assumed to be polynomials by writing $D : \text{polynomial} \to \text{polynomial}$.

Some authors like to use different kinds of arrows for these two purposes. This will not be done in this book. Notation becomes an obstacle rather than an aid to learning when it becomes fussy and distinctions are made beyond what is necessary to avoid misunderstandings. It is surely clear in the paragraph above that 'D:polynomial→polynomial' is indicating the general nature of the input and output, and is certainly not defining a specific operation. On the other hand, '$D : P(x) \rightarrow P'(x)$' does tell us that if we have a particular input, $P(x)$, we have a definite output, its derivative $P'(x)$, and this defines the operation D once we know what kind of input is acceptable. We need the two statements together to convey the full meaning of D, the operation of differentiation as applied to polynomials.

There is an aspect of notation in which students are bound to encounter some confusion, since there seem to be differences not only between past and present usage, but also between the conventions used to-today in different branches or at different levels of mathematics.

In the past all mathematicians used words such as 'the function x^2' or 'the function $f(x)$'. For instance, in connection with work on the remainder theorem sentences would be found such as 'when x^2 is divided by $x - a$ the remainder is a^2'. It is clear here that x^2 and a^2 must be interpreted differently. There is no doubt at all about what a^2 denotes: a stands for any number you like, say 9, and a^2 stands for its square, 81. If we assumed that x^2 was to be interpreted along the same lines, we could choose a number for x, say 10. Substituting these numbers, the statement above would become 'when 100 is divided by 1 the remainder is 81' which is surely not what the author intended. Rather the intention was to consider what happens when the polynomial x^2 is divided by the polynomial $x - a$ according to the usual algorithm, and observing that the constant remainder is given by a^2. If we were doing long division with a computer the symbol x would not reach the computer at all. We would have a programme in which the quadratic $bx^2 + cx + d$ was represented by its coefficients (b, c, d). The computer would be asked to divide $(1, 0, 0)$ by $(0, 1, -a)$ and report the remainder; it would respond by giving the single number a^2.

The distinction was not always very clearly expressed, but students of the older algebra books came to realize that letters from the end of the alphabet were to be interpreted differently from those at the beginning; in a context such as the above a represented a number, while x could not be so replaced, but was a sign used to indicate the polynomials involved in the division.

Mathematicians working at an advanced level in functional

analysis found this convention totally unacceptable. They replaced the older phrase 'the function x^2' by 'the function $x \to x^2$' or a longer phrase such as 'the function f defined by $f(x) = x^2$'. They would refer to $f(a)$ as 'the value of the function at a' and they did not mind where letters came in the alphabet; for them $f(x)$ represented the number that was the output corresponding to the input x—that is to say, it specified a particular number, not a function. Of course, by considering different values of x in turn and observing the corresponding numbers $f(x)$ we could see what the function f was like; this is the procedure we use when drawing a graph.

Algebraists sometimes use the symbol $f(x)$ in a way that resembles the older custom. For instance in his *Modern algebra* (a book of some repute) Van der Waerden does not hesitate to say 'Let $f(x)$ be a polynomial'. But then in algebra one is not always thinking of a polynomial as defining a function; one may be thinking of it simply as an object, and saying for instance that when certain rules of multiplication are applied to the objects $x + 1$ and $x - 1$ the result is the object $x^2 - 1$.

Earlier in this section reference was made to 'the polynomial $P(x)$' and to its derivative $P'(x)$. This is akin to Van der Waerden's symbolism rather than to the symbols for functions in functional analysis.

The older notation may be preferable for elementary work. It is probably easier for an engineering apprentice to learn that differentiating x^2 gives $2x$ than to cope with the more modern form, that differentiating the function $x \to x^2$ gives the function $x \to 2x$. We might show the latter diagrammatically like this;—

$$x \to x^2$$
$$D\downarrow$$
$$x \to 2x$$

This looks more complicated but it does reflect more accurately what is happening. Suppose we wish to differentiate some function which is given not by a formula but by numbers in a table. What we feed into the computer is not a single number, but the table, showing various values of x and the values $f(x)$ corresponding to these. Thus the symbol $x \to f(x)$ is quite an appropriate symbol for the function that constitutes the input. The computer then carries out some algorithm that approximately represents differentiation;

this corresponds to the operation D. The output again is not a single number, but a table for the derived function, $x \rightarrow f'(x)$.

Generally speaking, the modern notation is superior for functional analysis and for advanced analysis generally. However, as was stressed at the beginning of the book, numerical analysis spans the ages, it combines ancient and modern. It may well be that the majority of books on numerical analysis use the traditional notation. It is difficult to keep strictly to the modern style in such work, particularly when something simple is being discussed. For instance, we shall meet a symbol, $\|f - g\|$. This is a number, which measures how far apart two functions, f and g, are. So far, we have used modern notation. Now suppose we come to a situation where the traditionalist would happily write '$\|f(x) - x\|$ is small'. This means that the graph of $y = f(x)$ is close to the graph of $y = x$. The approximate equality of two functions, not of two numbers, is asserted. In modern notation, we have to experiment with phrases such as '$\|f - g\|$ is small, where g is the function $x \rightarrow x$', which is much more wordy and makes the whole thing sound as if it were rather difficult. The convention used in this book is to keep as close as is reasonable to the modern usage.

2.3. Continuity and distance

Continuity is a matter of life and death for numerical analysis. It arises even in simple calculations such as finding the value of π^2. We cannot take the exact value of π as our input, but only a close approximation to it. We hope this will give us a good approximation to π^2, and in fact, as squaring is a continuous operation, our hope will be realized. Informally stated, continuity means that a small change in input will lead to a small change in output.

That continuity should not be taken for granted, even in simple arithmetical operations, is shown by the following example.

Exercise. Solving a certain equation by Newton's method leads to the iteration with

$$x_{n+1} = \frac{2x_n^3}{3(x_n^2 - 1)}$$

Iteration is continued until x_{n+1} and x_n differ by less than some agreed amount, say 0.000 001. Find the numbers that result when the following values for the initial x_0 are taken; (a) 0.81, (b) 0.78, (c) 0.779, (d) 0.775, (e) 0.7747, (f) 0.77462, (g) 0.7746, (h) 0.77.

An appeal to continuity is involved in almost every computation. If we wish to determine $\int_0^x \phi(t)\,dt$, where ϕ is a function that cannot be integrated in terms of elementary functions, we use some quadrature procedure which depends on replacing ϕ by some function that can be integrated easily, and that approximates to ϕ. This approximating function may have a broken line graph, as in the trapezium rule, or it may be a polynomial, or it may be formed by using different polynomials in different intervals. The success of such procedures depends on the fact that integration is in fact a continuous operation in a sense that we shall make precise later. Incidentally, in this sense, the operation of differentiation is not continuous.

We might use iteration and approximate quadrature to solve an integral equation such as, for example,

$$f(x) = 1 + \int_0^1 \sin(x^2 + y^2) f(y)\,dy$$

and here the same problem confronts us; we have made a little change in the input to the problem, will this make only a little change in the output?

Indeed, in computing the question is not only the one just asked, but rather—how little must the change in the input be? Problems arise in computing that do not occur in pure mathematics, since computation is always to a certain number of decimal places, while in pure mathematics numbers are regarded as known exactly. For example, suppose a number, x, has to be found from the equations

$$(1+k)x + y = a$$
$$(1+2k)x + (1+k)y = b.$$

These give $k^2 x = (1+k)a - b$. From the viewpoint of pure mathematics, x depends continuously on a and b, provided $k \neq 0$. But suppose $k = 10^{-6}$. Any error in a or b will be multiplied by 10^{12} when x is calculated. If we want x correct to 6 places of decimals, we need a and b correct to 18 places, which may be quite impossible to obtain. Equations such as those just considered are called *ill-conditioned*.

In the early years of this century Maurice Fréchet set out to make a systematic study of functions T, with $g = Tf$, where f and g might not represent simply numbers but could be vectors, functions or other entities. He was guided in this work by analogies with the classical theory of functions of a real variable. One of the first questions he had to consider was—what do we mean, in this wider setting, by T being continuous?

Geometrical ideas arise naturally when such a question is considered. Classically, f is continuous if $f(a)$ and $f(b)$ are near to each other when a and b are near to each other. To find out how close a and b are, we measure the distance between them. For real numbers, this distance is $|a - b|$.

We may use the symbol $d(P, Q)$ as an abbreviation for 'the distance from P to Q'.

In order to generalize the idea of continuity Fréchet was led to define a *metric space* as any collection of objects for which a satisfactory definition of the distance between each pair of objects has been given. What does *satisfactory* mean?

It will be agreed that, if A, B, C, ... are points in 1, 2 or 3 dimensions, distance has the following properties;

(D.1) The distance from A to B is measured (in some suitable unit) by a real number $d(A, B)$.

(D.2) $d(A, B)$ is never negative.

(D.3) $d(A, B) = 0$ when A coincides with B and never in any other circumstances.

(D.4) The distance from A to B equals the distance from B to A.

(D.5) You cannot shorten your journey by breaking it. If you go from A to C and then from C to B, you have gone at least as far as from A to B direct. So, whatever A, B, C may be, $d(A, C) + d(C, B) \geqslant d(A, B)$.

This of course is a theorem of Euclid, that the sum of the lengths of two sides of a triangle is never less than the length of the third side. The inequality is referred to as *the triangle inequality*.

The five properties just listed will be referred to as *the axioms for distance*. They do not of course include everything that can be said about distance in our everyday geometry. For instance, Pythagoras' Theorem cannot be deduced from them. But if we examine the proofs of many theorems in classical analysis we find that only these properties are being used, and therefore the conclusions may be carried over to any system in which distance has been defined so that these axioms are satisfied.

Metric spaces are of great variety. If we imagine an insect crawling on the surface of a statue, and define $d(A, B)$ as the

3	3	3	3	3	3	3	4
3	2	2	2	2	2	3	4
3	2	1	1	1	2	3	4
3	2	1	♔	1	2	3	4
3	2	1	1	1	2	3	4
3	2	2	2	2	2	3	4
3	3	3	3	3	3	3	4
4	4	4	4	4	4	4	4

FIG. 1

distance the insect walks in going from A to B by *the most direct route* (or a most direct route, if there are alternatives), it is easily verified that $d(A, B)$ has all the properties listed above.

In chess, the king can move to any of the 8 squares adjacent to the square he is on. We could define the distance between any two squares as the number of moves a king is compelled to make to get from one to the other. The numbers on the squares in Figure 1 indicate the distances of the squares from the king's present position.

We carry over to metric spaces some definitions that are natural in Euclidean geometry.

Definition. $S(a, r)$ denotes the sphere with centre a and radius r. It consists of all the points, p, that are at a distance r from the centre a. In set notation, $S(a, r) = \{p : d(p, a) = r\}$.

Definition. $B(a, r)$ denotes the open ball with centre a and radius r. It consists of all the points p that are a distance *less than* r from the centre a. So $B(a, r) = \{p : d(p, a) < r\}$.

Definition. $\bar{B}(a, r)$ denotes the closed ball with centre a and radius r. It contains all the points that are at a distance r, or less, from a. So $\bar{B}(a, r) = \{p : d(p, a) \leqslant r\}$.

In the chess diagram, if k denotes the square where the king is shown, $S(k, 3)$ consists of all the points marked 3. $B(k, 3)$ consists of all the points marked 2 or 1, together with k. $\bar{B}(k, 3)$ consists of those with 3, 2, 1 or k.

It will be observed that in this metric space the 'sphere' $S(k, 3)$ appears to us as a square, while the sphere $S(k, 4)$, consisting of all the points marked 4, is even stranger, having the shape of an L. There is no need to be disturbed by these novelties. We simply have to abide by the generalized definitions and draw the logical consequences. It is perhaps surprising that the chess king's metric is closely related to a metric space of importance for numerical analysis, as will appear in a moment.

There is another way in which the word 'sphere' may appear peculiar. In our familiar plane, Euclidean space of two dimensions, $S(a, r)$ is the circle, centre a and radius r, and we do not normally apply the term 'sphere' to it. The reason for our present terminology is that we wish to have a word that will apply in space of n dimensions, regardless of whether n is 1, 2, 3, 4, 5, 17 or whatever. There is no word in everyday speech for $S(a, r)$ in four or more dimensions, so the word 'sphere', originally associated with three dimensions, has been chosen to stand for this general concept. So for $n = 2$ a 'sphere' is a circle, and for $n = 1$ simply a pair of points.

Definitions of distance. When we are dealing with some collection of objects, it often happens that there are several different ways of defining distance, and we need to choose the definition best suited to our immediate purpose. Suppose for instance we calculate a pair of numbers and errors x and y are involved in our results. What definition of distance gives the best measure of how far away we are from the truth? It depends on the circumstances. If our numbers are the co-ordinates of a point in the plane, it would be natural to use the Euclidean distance $\sqrt{(x^2 + y^2)}$. If our numbers were entries in a table and we wanted to say to how many places of decimals our results were reliable, it would be natural to examine the magnitude of the errors, x and y, and give the larger of these. This would be written, distance $= \max\{|x|, |y|\}$. Again it might happen that a further calculation was due to be made in which the numbers would be added. In this case the errors also might be added (if they were in the same direction) and an error of magnitude $|x| + |y|$ could result. This number, $|x| + |y|$, would then be the most useful definition of distance from the correct values. Figure 2 shows the shapes of the 'spheres' that would result from these three definitions.

$$d(x,y) = \sqrt{(x^2 + y^2)} \qquad\qquad d(x,y) \qquad\qquad\qquad d(x,y) = |x| + |y|$$
$$\qquad\qquad\qquad\qquad\qquad = \max\ \{|x|,|y|\}$$

Fɪɢ. 2

It will be noticed that the second diagram resembles that for the king's move metric. The third diagram has some relation to distances in city streets. Imagine that the lines on squared graph paper represent streets and that you can get from one point to another only by walking along these lines. The places you could reach by a fixed amount of walking from a given street corner would lie on a tilted square like that in the diagram for $d(x, y) = |x| + |y|$.

The corresponding possibilities arise when we have n numbers with errors, $x_1, x_2, \ldots x_n$. We may use $\sqrt{(x_1^2 + x_2^2 + \ldots + x_n^2)}$, which is known as the ℓ_2 distance, or $\max\{|x_1|, |x_2|, \ldots |x_n|\}$ called the ℓ_∞ distance, or finally $|x_1| + |x_2| + \ldots + |x_n|$, the ℓ_1 distance. The reasons for the subscripts 2, ∞ and 1 will appear later when we consider the ℓ_p distance.

It should be emphasised that in the discussion above the algebraic quantities have represented *errors*. Thus, in the two-dimensional case considered first, $x = x_1 - x_0$ where x_1 is the computed value and x_0 the true value. Similarly $y = y_1 - y_0$. Thus if d denotes the distance from (x_0, y_0) to (x_1, y_1), the formulas implied by our three definitions of distance above are;—

$$\ell_2 \text{ definition,} \quad d^2 = (x_1 - x_0)^2 + (y_1 - y_0)^2$$
$$\ell_\infty \text{ definition,} \quad d = \max\{|x_1 - x_0|, |y_1 - y_0|\}$$
$$\ell_1 \text{ definition,} \quad d = |x_1 - x_0| + |y_1 - y_0|.$$

Exercises and points to consider

1. Prove that the ℓ_1 and ℓ_∞ definitions of distance in two dimensions do satisfy the distance axioms (1) to (5).

2. For the ℓ_2 definition of distance in two dimensions, properties (1) to (4) are easily seen to be true, and we know property (5) as a theorem in geometry. It must be possible to prove property (5) from the definition above purely by means of algebra, but it is not obvious how to do so. Consider this problem.

3. A signal sent by telegraph or fed into a computer may be represented by a series of noughts and ones. 0 indicating *no pulse* and 1 denoting *a pulse*. In the study of errors that may arise in transmission the Hamming distance is defined as the number of errors in the first signal that would be required to turn it into the second signal. For instance the distance between the signals

$$
\begin{array}{ccccc}
* & & * & * & \\
1 & 0 & 1 & 0 & 1 \\
0 & 0 & 0 & 1 & 1
\end{array}
$$

is 3; the signals differ at the points marked *. We will assume that all signals are of length 5 as in the example above.

(a) Does the Hamming definition satisfy the axioms of distance and so define a metric space of signals?

(b) Let $a = 10101$. How many signals are there in $S(a, 2)$? in $\bar{B}(a, 2)$?

(c) What is $S(a, 5)$?

4. What point (x, y) on the line with equation $2x + y = 5$ is nearest to the origin in the metric ℓ_∞?—in metric ℓ_2?—in metric ℓ_1? What are the distances in these three metrics? When these distances are arranged in order of magnitude does there appear to be any connection with the order of the numbers $1, 2, \infty$?

5. If the surface of the earth is regarded as a perfect sphere of radius R, distances are measured by the shortest routes on the *surface* (not by chords through the earth), and N represents the North Pole, what are the usual geographical names of the following?—

(a) $S(N, \pi R/2)$.   (b) $B(N, \pi R/2)$ (c) $\bar{B}(N, \pi R)$.

(d) $B(N, \pi R)$.   (e) $S(N, \pi R)$.

2.4. Limits and convergence

In connection with continuity we considered whether a slight error in our input might destroy the value of our answer. If our computation involves iteration there is also the question whether we shall get any answer at all. We might instruct the computer to calculate a sequence of points, (x_n, y_n), and stop when both $x_{n+1} - x_n$ and $y_{n+1} - y_n$ are less than 0.001 in magnitude. It is possible that the computer might run for ever without this happening.

It is not intended here to suggest that the question is finished if it *does* happen. Here too ill-conditioning can have disturbing effects. For instance, if we have

$$x_{n+1} = 0.409\ 651\ 567 x_n + 0.647\ 202\ 229 y_n - 0.056\ 853\ 797$$

$$y_{n+1} = 0.647\ 202\ 229 x_n + 0.290\ 248\ 433 y_n + 0.062\ 549\ 338$$

and take $x_0 = 2$, $y_0 = 1.912$, we get the table

n	x_n	y_n
0	2.000 000 000	1.912 000 000
1	1.999 900 000	1.911 908 800
2	1.999 800 010	1.911 817 609
3	1.999 700 030	1.911 726 427

In each column, the change from one row to the next is of the order of 0.0001, but this does not mean we are anywhere near the situation in which input and output are equal. That happens when $x = 1$, $y = 1$, towards which the iteration is proceeding at a snail's pace. In fact, if we consider the errors, $x_n - 1$ and $y_n - 1$, we find that, as we go from one row to the next, each of these is multiplied by 0.9999. The smallness of the change is due to the nature of the iteration, not to our being near the solution of the equation from which this iteration would arise.

Accordingly, even when the computer stops, further checks are necessary before we can be sure we have found a solution of our problem. Having recognized the existence of this loophole, we return to the main theme of this section, the problem of whether the values calculated by the computer do eventually settle down in some small neighbourhood.

This question is closely related to the questions in pure mathematics, does x_n tend to a limit when $n \to \infty$?—does $x(t)$, depending on a continuous variable t, tend to a limit as $t \to \infty$?

Classical mathematics deals with these questions by a two-stage process. First, there is a definition which explains what a limit means, but is not much use in dealing with particular problems. Secondly, a condition of more practical value is derived from this definition. We shall briefly review these.

The theory of limits presupposes an opponent who can demand arbitrarily small tolerances. If we assert that $e^{-t} \sin t \to 0$ as $t \to \infty$, the opponent may require us to prove that there is a time t_0 after which $e^{-t} \sin t$ stays between $-0.000\,001$ and $+0.000\,001$. The situation is represented in Figure 3. The tolerances demanded by the opponent create a kind of tube which the graph must enter and remain inside. It is of course easy to meet the opponent's demand,

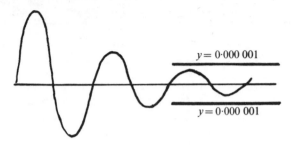

$y = 0\cdot000\,001$

$y = 0\cdot000\,001$

FIG. 3

since $|\sin t| \leqslant 1$ and $t \geqslant 14$ ensures $e^t \geqslant 1\,000\,000$. The argument can be adapted without difficulty to any other tolerance the opponent may choose.

If we were interested not in a continuous function but in a sequence, $x_n = e^{-n} \sin n$, we would not need to use the whole graph in Figure 3, but only to pick out the points on it corresponding to $t = n$, with $n = 1, 2, 3, \ldots$.

A function is defined to have a stated limit if, whatever tolerance the opponent may demand, the values of the function *arrive at and stay within* that tolerance.

The weakness of this test for practical purposes is evident. To apply it, we have to start with a stated limit. But if we are performing an iteration in order to compute some number, the whole point of the operation is that we do not know in advance what that number is. We are relying on the iteration itself to tell us the approximate value of the limit. What we need is a test that will show whether a sequence tends to *some* limit.

Such a test was provided by Cauchy about a century and a half ago. The idea is that if the successive numbers $x_n, x_{n+1}, x_{n+2} \ldots$ are getting close to some limit, they must be getting close to each other. To take a very simple example, consider the sequence 1, 1.1, 1.11, 1.111, ... that we had in connection with Achilles and the tortoise. The difference between the approximation 1.111 and the later approximation 1.111 111 is 0.000 111, which is less than 0.0002. If we had taken the later approximation further on, we would have had more ones in the figure for the difference, 0.000 11 ... 11, but the difference would still be less than 0.0002. In the same way, by taking 1.111 111 as our first approximation, we can be sure that no later approximation will modify it by as much as 0.000 000 2. In

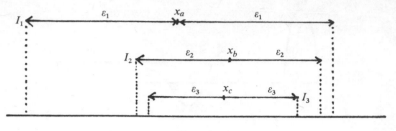

FIG. 4

fact, we can ensure any degree of accuracy we may wish. Note that in this argument we nowhere mention or need to know the number $\frac{10}{9}$ to which in fact the sequence tends.

The exact statement of this principle is as follows. Suppose the opponent produces any positive number. We will follow tradition and call his number epsilon, ε. It is now up to us to find an approximation x_n, such that no later approximation differs from x_n by more than ε. We can express 'a later approximation' by letting p stand for any of the numbers $1, 2, 3, \ldots$. Then every later approximation is of the form x_{n+p}. The requirement thus is that $|x_n - x_{n+p}| \leqslant \varepsilon$ for the n we have chosen and *for every natural number p*. That is to say, the entire 'tail' of the sequence, from x_n on, must lie in the interval $[x_n - \varepsilon, x_n + \varepsilon]$.

What this means graphically is shown in Figure 4. The opponent first produces ε_1 and we respond by taking $n = a$. This means that all the terms of the sequence from x_a on lie in the interval $[x_a - \varepsilon_1, x_a + \varepsilon_1]$. Call this interval I_1. If the opponent now produces a smaller number ε_2, we shall choose a later term x_b so that all terms from x_b on lie in the smaller interval $[x_b - \varepsilon_2, x_b + \varepsilon_2]$. Now x_b certainly lies in the interval I_1 since all the terms later than x_a do so. In Figure 4 we have shown the interval $[x_b - \varepsilon_2, x_b + \varepsilon_2]$ lying entirely in the interval I_1. It may be objected, with reason, that one or other end of $[x_b - \varepsilon_2, x_b + \varepsilon_2]$ may lie outside I_1. (They cannot both lie outside as the interval in question is not as long as I_1.) If this should happen, it does not matter. No term after x_a lies outside I_1, so certainly no term after x_b lies outside. Accordingly if we define our second interval, I_2, as the interval that is common to I_1 and $[x_b - \varepsilon_2, x_b + \varepsilon_2]$, we can be certain of the following things, (a) all the terms from x_b on lie in I_2, (b) I_2 is contained in I_1, (c) the length of I_2 is $2\varepsilon_2$ *or less*.

We continue in this way. When the opponent produces a still smaller number ε_3, we find a term x_c, so that all the terms from x_c on are contained in an interval I_3, which has length not more than $2\varepsilon_3$, and which lies in I_2.

We are assuming that, however small a number the opponent chooses, we can find a response that meets the requirements. If this is so, the sequence must tend to a limit. This limit is a point that lies in all the intervals I_1, I_2, I_3, I_4 ... as the construction above is carried out indefinitely, the opponent choosing a sequence of numbers that tend to 0.

It is clear that there cannot be two distinct points that lie in all the intervals I_n. For if the distance between two points is k, there will be a stage at which $2\varepsilon_n < k$, and the corresponding I_n, with a length less than $2\varepsilon_n$, will not be long enough to reach from one point to the other.

Somewhat deeper considerations are involved in the question of whether there must be a point common to all these intervals. If there were no numbers except the rational numbers, the intersection of all the intervals could be empty. For instance consider the intervals $[1, 2]$, $[\frac{3}{2}, \frac{5}{3}]$, $[\frac{8}{5}, \frac{13}{8}]$, $[\frac{21}{13}, \frac{34}{21}]$ ... the numbers appearing here being taken from the Fibonacci sequence 1, 2, 3, 5, 8, 13, 21 ... in which each number is the sum of the two previous numbers. These intervals close in on the number $(1 + \sqrt{5})/2$ which is irrational, and so would not appear if we could have a line on which only the rational numbers were represented by points.

Accordingly, to prove that intervals closing in as described earlier are bound to contain a point, we have to develop a theory (such as Dedekind sections) to explain what irrational numbers are, and from this theory obtain the desired theorem.

The process described above, with intervals closing in, is like a trap in which we try to catch the limit of a sequence. A system, like the rational numbers, in which the trap can close simply on emptiness, is called *incomplete*. It is full of holes. A system, like the real numbers, in which the trap is sure to catch a point (or a number, whichever you like to call it) is *complete*.

A sequence that passes the test we have been describing is called a *Cauchy sequence*. We repeat the condition it satisfies; however small a positive number, ε, may be, we can always find a number, n, such that $|x_n - x_{n+p}| \leqslant \varepsilon$ for every number p chosen from 1, 2, 3, ...

This condition expresses that the sequence is settling down in a reasonable manner. It is not wandering all over the place like 1, −2, 3, −4, 5, −6 . . . which we certainly do not expect to have a limit. It is not even vacillating between two numbers like the sequence 1, −1, 1, −1, ... Nor again is it doing what is described as tending to an infinite limit, like 1, 2, 3, 4, 5, 6 . . . . Infinity finds a place in some schemes, but it is not accepted as a member of the real number system.

It has taken some time to describe in detail the idea of a Cauchy sequence of real numbers. This however seems to be justified since this idea, and its generalization to other systems, continually appears in connection with tests for convergence. Fortunately, the generalization to metric spaces is immediate and does not involve any more work.

In the condition which has just been repeated, $|x_n - x_{n+p}| \leq \varepsilon$, the expression $|x_n - x_{n+p}|$ represents the distance between the points x_n and x_{n+p} on the real line. All we have to do to obtain the condition for a Cauchy sequence in a metric space is to replace $|x_n - x_{n+p}|$ by the expression $d(x_n, x_{n+p})$.

Thus if $x_1, x_2, x_3 \ldots$ is a sequence in some metric space, it will be a Cauchy sequence if, however small the positive ε chosen by the opponent may be, we can find an x_n in the sequence such that every term x_{n+p} coming after x_n satisfies $d(x_n, x_{n+p}) \leq \varepsilon$. This means that all the terms coming after x_n, what we may call the tail of the sequence, lies in $\bar{B}(x_n, \varepsilon)$, the closed ball with centre x_n and radius ε.

Figure 5 illustrates the situation for a Cauchy sequence in two dimensions. All the terms from x_a on lie in or on the circle with centre x_a. There is a later term, x_b, such that all the terms after x_b lie in the smaller circle with centre x_b. Similarly there is a still smaller circle, x_c, and so it continues, the circles narrowing down and identifying a single point, the limit of the sequence. While this figure relates to two dimensions only, it is very helpful for visualizing the meaning of a Cauchy sequence in *any* metric space.

A sequence of regions, each lying in the previous one, is called *nested*. Thus in Figure 5 we have a sequence of nested closed balls.

When we were dealing with Cauchy sequences in one dimension, that is with sequences of real numbers, we had to consider the possibility that part of the interval centred on x_b might lie outside the interval centred on x_a. We decided to ignore any such part, since no points could lie in it. The same possibility arises here. Part,

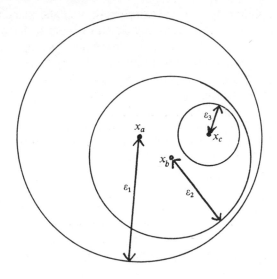

Fig. 5

though not all, of the circle centred on x_b may lie outside the circle with centre x_a. If this happens, as before, we ignore the part that is outside. We do the same if, at any later stage, part of one closed ball lies outside the previous closed balls. In this way we still obtain a sequence of nested regions that shrink towards a single point, though the shape of the regions now may be more complicated. However as each region is part of a ball, and as the radius of the ball tends to nought as we go down the sequence, the enclosed points have no option but to tend to a limit, provided always that we are dealing with a complete space and not a perforated one like the rational numbers.

We have dealt first with the Cauchy condition for any metric space, that is, with the practical condition used to decide whether a sequence tends to a limit. We must also deal with the basic theoretical definition, to show what we mean by 'tending to a limit' in any metric space. This again gives no trouble. For real numbers, $x_n \to l$ means that $|x_n - l|$, the size of the difference between x_n and l, becomes and stays as small as anyone may require.

All we have to do to get our definition in an arbitrary metric space is to replace $|x_n - l|$, which measures a distance on the real line, by $d(x_n, l)$. Thus $x_n \to l$ if the distance of x_n from l becomes

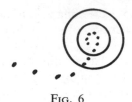

FIG. 6

and remains as small as anyone may require. In geometrical terms this means that if we take a ball with centre l, however small the radius may be, the sequence will enter and stay inside this ball. This is illustrated in Figure 6.

The formal statement of this is contained in the following definition.

Definition. In any metric space, 'x_n tends to the limit l', has the following meaning; for any $\varepsilon > 0$, there exists a number N such that $d(x_n, l) \leqslant \varepsilon$ for every n satisfying $n \geqslant N$.

If you examine this statement, you will see that it is equivalent to saying that $d(x_n, l) \to 0$. Thus there is a shorter, and very reasonable sounding definition of tending to a limit in a metric space. We say that $x_n \to l$ if the distance of x_n from l tends to nought.

2.5. The role of the distance axioms

An essential feature of the theory of limits is that a convergent sequence can have only one limit. If we visualize points in a plane, we will readily agree that points which are steadily approaching a point, P, cannot at the same time be steadily approaching a different point, Q. But we cannot feel so sure about the corresponding theorem for points in an arbitrary metric space. There exist a great variety of metric spaces; our picture of a matric space is not clear cut and definite but somewhat shadowy. There is only one way to achieve certainty. All we know about a metric space is that it carries a satisfactory definition of distance, satisfying the axioms D.1 to D.5 given earlier. The theorem, if true, must follow from these axioms.

Suppose then, if it is possible, that in some metric space a sequence (x_n) tends both to P and to Q. This means that, if ε is any positive number, there is N such that all members of the sequence, from x_N on, are at a distance from P not exceeding ε. Similarly there is M such that all members of the sequence, from x_M on, are at a distance from Q not exceeding ε. It is easy to pick a member of the sequence that satisfies both conditions; x_{M+N} will do. So $d(x_{M+N}, P) \leqslant \varepsilon$ and $d(x_{M+N}, Q) \leqslant \varepsilon$. At this stage our geometrical imagination, even though based only on the everyday world of 3 dimensions, gives us a strong nudge as to what we should do next. If x_{M+N} is so close both to P and to Q, P and Q must be pretty close to each other. The only axiom that can possibly help to formalize this belief is D.5, the triangle inequality. The journey from P to Q cannot be shortened by breaking it at x_{M+N}. That is

$$d(P, x_{M+N}) + d(x_{M+N}, Q) \geqslant d(P, Q).$$

We have to use D.4, which says the distance is the same whether you are going or returning, to replace $d(P, x_{M+N})$ in the inequality above by $d(x_{M+N}, P)$. We then see that each distance on the left-hand side of the inequality is ε at most. We thus find $d(P, Q) \leqslant 2\varepsilon$, where ε is any positive number you like to choose. Axiom D.2 tells us that a distance is never negative, so it follows that $d(P, Q) = 0$. D.3 tells us that a distance is nought only when the points in question coincide. So P and Q are identical.

This disposes of the possibility of there being two *distinct* points P and Q, both being a limit of the same convergent sequence.

It will be noticed that the argument has used all the axioms of distance. We have not explicitly used D.1, that distance is measured by a real number, but clearly that has been understood throughout the argument.

This proof also illustrates the fact that the five simple axioms, D.1 to D.5, have caught enough of the essence of the idea of distance to allow us to prove for the general metric space many theorems which in their traditional setting appear plausible on geometrical grounds.

2.6. Vectors

Mathematics often follows the axiomatic route to generality. This is related to an experience many students must have had. They have been confronted with a problem of a new type. They do not have any feeling for its nature but by long and complicated calculations they succeed in reaching a solution. Later, perhaps by examining the answer itself or in the light of further knowledge, they realize that a much simpler treatment is possible. They have thus arrived at an idea or a principle that quite possibly can be used to solve other problems.

A similar thing happens in the progress of mathematics. The first proof of a theorem is often clumsy and unenlightening. Perhaps over the course of decades a really good proof is reached—one that does not go into unnecessary details, but reveals the heart of the matter. Then, in order to detach a generalization from the particular instance in which it arose, the proof is analysed. What assumptions does it make? If these assumptions are listed and we meet other objects for which these assumptions are true, we know without further ado that the theorem will hold for those objects also.

Axiomatization—this process of analysing underlying assumptions—also leads to the dissection of a subject. If ten assumptions or axioms are needed, we may enquire what theorems follow from each axiom by itself, from each possible pair of axioms, and so forth. Needless to say, only combinations of axioms that seem significant are studied; otherwise the task would be of unbearable complexity. Thus theorems may be proved that hold for any group. Further theorems result if we assume the group to be commutative, others if we suppose it to be finite, still more if we suppose it both finite and commutative.

It is a defect of many expositions of vector theory that—at any rate from the axiomatic viewpoint—they present the student with an incredible jumble. Vector sum, vector product, scalar (or 'dot') product, and the length of a vector appear in the first chapter as though they all stood on an equal footing, which is far from being the case, axiomatically speaking. Such a treatment may be perfectly satisfactory for a limited objective, such as the study of elementary mechanics or physics. It is not helpful for anyone who wishes to go on to more advanced mathematics.

As we shall soon see, one of the basic ideas of functional analysis is to regard functions as vectors. This is to be understood not only as applying to functions in the sense of classical analysis, but in the wider sense discussed earlier; a function is defined by any process for which to any input there corresponds one definite output.

For instance, if p denotes any polynomial, let Q denote the operation of squaring it, and D that of differentiating it. Both Q and D qualify as functions, polynomials→polynomials, for there is no doubt what the square of a polynomial is, and the same for the derivative. So we shall want to treat Q and D as vectors.

In elementary vector theory, if u and v are two vectors, we can form their scalar product, $u \cdot v$, and we can also form a linear combination such as $10u + 3v$. How do these things fare when we come to consider Q and D?

The scalar product, $u \cdot v$, is related to the lengths of u and v, and to the angle between them. There may be some sensible way of defining the angle between Q, squaring, and D, differentiating, but at any rate it is not obvious, and it may not exist at all. So we are going to regard scalar product as an optional feature; some vector systems may have it, others may not.

It is entirely otherwise with linear combinations. There is no difficulty in defining $10Q + 3D$. To define it we have to say what this operation does to any polynomial, p. We do not have to look far for an interpretation of this symbol. It is very natural and reasonable to write

$$(10Q + 3D)p = 10(Qp) + 3(Dp) = 10p^2 + 3p'.$$

This will be regarded as the essential property of a vector space— that we can form linear combinations of its elements.

2.7. Pure vector theory

By pure vector theory we shall understand what results from axioms that are just sufficient to guarantee that such linear combinations exist and that they behave in a reasonable manner.

For the existence of linear combinations, it is sufficient to assume that the sum $u + v$ is defined for any two elements, u and v, of a vector space, and that ku is defined for any u in the space and for any real number k. Thus our only fundamental operations will be addition of vectors and the multiplication of a vector by a number.

We still have to spell out what we mean by these operations behaving in a reasonable manner. We shall shortly do this formally by listing ten axioms. It is possible and desirable to grasp the meaning of these ten axioms by a single mental act; this can be done with the help of a rather childish and simple-minded model of a vector space.

Somewhere, not far from kindergarten level, children may learn to associate the idea of $2 + 3$ with that of 2 cats and 3 cats. There is a long road from this to an appreciation of the real number system. The child has to extend the idea of number to fractions, to negative numbers, to irrational and even transcendental numbers. Yet all of this grows in a fairly natural and inevitable way from the ideas of addition and multiplication as originally understood in relation to collections of cats.

It is conceivable (though presumably not desirable) that young children could be led to associate the symbol $(2, 3)$ with the idea of 2 cats and 3 dogs. They could then arrive, by a purely experimental procedure, at results such as $10 \times (2, 3) = (20, 30)$ and $(2, 3) + (4, 5) = (6, 8)$. If they were taught to plot points on squared graph paper, they could observe that multiplication by 10 had the effect of a dilation and that addition involved points which, taken together with $(0, 0)$, formed a parallelogram.

However such plotting would be a distorting graphical device since it introduces properties that do not exist in the original situation. For instance, on squared paper the lines joining the origin to the points $(1, 0)$ and $(0, 1)$ are perpendicular. But these points represent a cat and a dog, and there is no obvious reason why we should regard a cat as perpendicular to a dog, or for that matter as making any other angle with it. Again, an arrow joining the origin to the point $(3, 4)$ is the vector that signifies 3 cats and 4 dogs. The length of this arrow is 5, and it is hard to imagine any sense in which 5 can be regarded as measuring the magnitude of the collection, 3 cats and 4 dogs.

Now of course the idea just described is only a starting point towards the understanding of a vector space, just as counting cats is only a beginning towards the understanding of the real number system. It would have to be developed in much the same way, by allowing the x and y in (x, y) to become fractional, negative and so forth. But this very primitive sketch of a vector system does have three points in its favour. First, it illustrates the possibility of pure vector theory, a theory in which angles and lengths are totally irrelevant, the only relevant ideas being the operations used to form $u + v$ and ku. Second, it provides a model that is helpful when we are discussing the formal properties of these operations. The ten axioms for a vector space can be summed up as requiring the formal properties that apply in the cat-and-dog system, suitably extended by the introduction of fractional and negative quantities. Third, it removes the psychological barrier at three dimensions. If vector spaces are first introduced in physical or geometrical terms, a feeling is likely to arise that there is something unreal or unfair about vector spaces of four or more dimensions. On the other hand, if we are discussing collections of animals, there is no reason to accept the idea of two or three species being involved, and reject the idea that there might be seven, or 19 or 243 kinds of animals.

In speaking above of the formal properties of operations, we have very general properties in mind. They will be related to such questions as, 'Does the order matter when vectors are being added?' Nothing at all will be said to restrict the kind of object that can qualify as a vector; the restrictions will simply be on the operations to be performed. Indirectly perhaps the difficulty of defining such operations may disqualify certain types of object from being considered as vectors.

2.8. Axioms of a vector space

A vector space then involves certain things, which will be referred to as *vectors*, and certain operations on these. Any such system will be accepted as a vector space if it meets the ten conditions listed below. In these conditions, letters from the end of the alphabet, such as u, v, w, represent vectors, and those from the beginning of the alphabet, such as a, b, c, represent numbers.

(V.1) Corresponding to any two vectors, u and v, there is a definite vector, y, which is their sum; $y = u + v$.
(V.2) $u + v = v + u$.
(V.3) $u + (v + w) = (u + v) + w$.
(V.4) There is a certain vector, written 0 and called the zero vector, such that for any vector u we have $u + 0 = u$.
(V.5) Whatever the vectors u and v may be, there is exactly one vector z for which $u + z = v$. This vector z is denoted by $v - u$.
(V.6) For any number a and any vector v there is a definite vector av.
(V.7) For any number a, and any vectors u and v

$$a(u + v) = au + av.$$

(V.8) For any numbers a, b and any vector v

$$av + bv = (a + b)v.$$

(V.9) For any numbers a, b and any vector v

$$a(bv) = (ab)v.$$

(V.10) Multiplying any vector v by the number 1 gives v.

The algebraic symbolism here may obscure the fact that all these statements embody rules of operation that we perform without an instant's thought when we are actually making a calculation. For instance V.9 relates to a question such as 'What is 3 times $2v$?' We at once write $6v$, since $6 = 3 \times 2$. What we do—quite automatically and unconsciously—is to detach the 2 from the v, multiply this 2 by 3 to get 6, and then write the 6 in front of v. The brackets in V.9 express this procedure concisely—perhaps so concisely that the meaning is not immediately appreciated.

Exercises

State which of the following are vector spaces, the operations of addition and multiplication being defined in the natural way. For those which are not vector spaces, state an axiom that fails to be satisfied.

1. The set of all quadratic expressions, $Q(x) = ax^2 + bx + c$.

2. All quadratic expressions, $Q(x)$, having $Q(0) = 0$.

3. All quadratics satisfying both $Q(0) = 0$ and $Q(1) = 0$.

4. All quadratics satisfying all three conditions, $Q(0) = 0$, $Q(1) = 0$, $Q(2) = 0$.

5. All the expressions $x^2 + bx + c$.

6. All polynomials with degree not exceeding 5.

7. All polynomials.

8. All functions defined on $[0, 1]$, with real values.

9. All bounded functions, $[0, 1] \to \mathbb{R}$. This means that for each function, f, there is a number, M, such that $|f(x)| \leq M$ for $0 \leq x \leq 1$.

10. All functions, $[0, 1] \to \mathbb{R}$, with the bound $M = 100$.

11. All continuous functions $[0, 1] \to \mathbb{R}$.

12. All continuous functions, $[0, 1] \to \mathbb{R}$, with $f(0.5) = 0$.

13. All continuous functions $[0, 1] \to \mathbb{R}$ with $f(0.5) = 2$.

14. All functions, f, with f, f', f'' continuous, $[0, \pi] \to \mathbb{R}$, with f satisfying the differential equation, $f''(x) + f(x) = 0$, and the end conditions $f(0) = f(\pi) = 0$.

3 Iteration and contraction mappings

3.1. Testing convergence

WE SAW earlier that the Cauchy test for convergence of a sequence was more practical than the original theoretical definition of a limit. However it may still appear very awkward to use. It depends on choosing some term, x_n, of a sequence and showing that all later terms lie within a sufficiently small distance from x_n. This suggests that we have to calculate the distance $d(x_n, x_{n+p})$ for $p = 0, 1, 2, 3 \ldots$ and check that all these distances are less than some prescribed number. If they are not, we must try some larger number for n and repeat the process. It looks as if an infinite amount of work is involved.

Fortunately a considerable simplification of the task is often possible. This is due to a principle which we may call the polygon inequality. In Figure 7, it is evidently shorter to go from A to D directly rather than via B and C. There would be equal distances in these routes if B and C happened to lie on the segment AD. Thus in the Euclidean plane we have $d(A, D) \leq d(A, B) + d(B, C) + d(C, D)$. This will be true for any metric space if we can prove it from the triangle inequality. This offers no difficulty. For we have $d(A, C) \leq d(A, B) + d(B, C)$ and $d(A, D) \leq d(A, C) + d(C, D)$, from which our result immediately follows.

It is clear this proof could be extended to polygons with any number of sides. In fact, the principle that you cannot shorten your journey by breaking it holds for any finite number of breaks, in any metric space. In this present section, however, we are concerned only with the familiar Euclidean space.

In Figure 7, if we let v_1, v_2 and v_3 denote the vectors AB, BC and CD respectively, then the vector AD is $v_1 + v_2 + v_3$. The inequality we obtained in connection with Figure 7 can now be written

$$\|v_1 + v_2 + v_3\| \leq \|v_1\| + \|v_2\| + \|v_3\|$$

and quite generally we have

$$\left\| \sum_1^n v_r \right\| \leq \sum_1^n \|v_r\|.$$

Here $\|v\|$ denotes the length of the vector v.

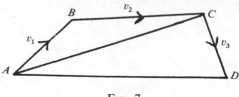

We are now going to consider the application of the polygon inequality to the convergence of sequences of points. We shall begin with very special cases and gradually widen the scope of the discussion as this chapter proceeds.

Suppose then we have a sequence of points, $P_0, P_1, P_2, \ldots$ in the Euclidean plane. We may think of P_0P_1, P_1P_2, P_2P_3, and so on, as links in a chain. Suppose we know the lengths of these links to be 1, 1/2, 1/4 and so on in geometrical progression, but we have no information at all about the directions in which the links are placed. What can we say about the distance of P_n from P_0? The length of the chain from P_0 to P_n is $1+(1/2)+(1/4)+ \ldots (1/2^{n-1})$. If the chain is stretched in a straight line, this number will give the distance $d(P_0, P_n)$; in any other circumstances the distance will be less. This is intuitively obvious, and formally is proved by the polygon inequality. However large n may be, the distance expressed by the sum above will never reach 2. Thus we can be certain that all points of the sequence lie inside the circle of radius 2 with centre P_0. In the notation for open balls, they lie in $B(P_0, 2)$.

If now we consider the links of the chain from P_1 to P_n, their total length is $(1/2)+(1/4)+ \ldots (1/2^{n-1})$ and so is less than 1. So all the points of the sequence that come after P_1 lie in the circle with centre P_1 and radius 1, that is, in $B(P_1, 1)$. Similarly we can see that the points coming after P_2 lie in $B(P_2, 1/2)$ and quite generally the points after P_k lie in $B(P_k, 1/2^{k-1})$. In fact we have a sequence of nested balls, with $1/2^{k-1}$, the radius of the ball with centre P_k, tending to 0 as $k \to \infty$. The sequence must converge; no choice of directions for the links can make it diverge, though of course that choice can alter the position of the point to which it converges.

You may have noticed that the argument above used open balls, such as $B(P_0, 2)$, while the statements about convergence in section 2.4 all involved closed balls. The reason for this is that all the points of the sequence lie *inside* the circle with centre P_0 and radius 2.

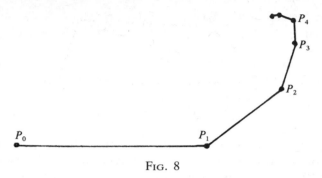

FIG. 8

However if all the links pointed in the same direction, the limit point they are approaching would lie *on* that circle. Accordingly, we have to use the closed balls $\bar{B}(P_k, 1/2^{k-1})$ rather than the open balls if we are trying to specify a succession of regions in which the limit point must lie.

The situation we have been considering is admittedly a very special one, yet it is in no sense unreal. Consider for instance the iteration

$$\left.\begin{array}{l} x_{n+1} = 0.4x_n - 0.3y_n + 1 \\ y_{n+1} = 0.3x_n + 0.4y_n \end{array}\right\} \tag{1}$$

with $x_0 = 0$, $y_0 = 0$ and (x_n, y_n) giving the point P_n. This sequence is shown in Figure 8. Each link in this chain is half as long as the previous one, and the direction of each link makes the same angle, about 37°, with that of the previous link.

The general argument, based on the lengths of the links alone, shows that the limit point lies within 0.000 007 7 of P_{18}. A more exact calculation, taking into account the actual directions of the links, does not lead to any very striking improvement of this estimate.

3.2. The idea of a contraction mapping

At the beginning of section 3.1, we studied convergence with the help of the polygon inequality expressed in the form

$$\left\| \sum_1^n v_r \right\| \le \sum_1^n \| v_r \|.$$

In that section $\|v\|$ indicated the length of a vector v in the Euclidean plane. We are now trying to make this approach available in a much wider setting.

As the inequality above involves the sum of vectors, we naturally assume that our generalization involves some system satisfying the vector axioms, V.1 to V.10. The new ingredient that we are bringing in is the generalization of the length of a vector. We shall use the same symbol, $\|v\|$, as before. Usually $\|v\|$ is referred to, not as the length, but as the norm of v. This is understandable. For example, as mentioned in chapter 2, the vector v may be a function, and it sounds strange, and perhaps misleading, to speak of the length of a function.

In view of the norm being a generalization of length, the following axioms are very reasonable;

N.1. With every vector, v, is associated a real number, $\|v\|$.
N.2. $\|v\|$ is never negative; $\|v\| \geqslant 0$.
N.3. $\|v\| = 0$ when and only when $v = 0$.
N.4. If k is any real number, $\|kv\| = |k|\,\|v\|$.

Figure 9 shows the construction for vector sum in the Euclidean plane. This figure involves a triangle and we embody the triangle inequality in the next axiom, N.5.

N.5. $\|u + v\| \leqslant \|u\| + \|v\|$.

In the Euclidean plane, the distance between (x_1, y_1) and (x_2, y_2) is given by $d^2 = (x_2 - x_1)^2 + (y_2 - y_1)^2$. Now if $u = (x_1, y_1)$ and $v = (x_2, y_2)$, the vector $(x_2 - x_1, y_2 - y_1)$ is $v - u$ and the formula above shows that $d = \|v - u\|$. We retain this result.

Definition. The distance between any two vectors u and v is defined to be $d(u, v) = \|v - u\|$.

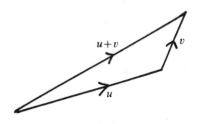

FIG. 9

By using the axioms N.1 to N.5 above, it can be shown that $d(u, v)$, so defined, satisfies the axioms for distance. The axioms D.1, D.2 and D.3 correspond closely to N.1, N.2 and N.3. To prove D.4, which requires us to show $d(u, w) = d(w, u)$, we use N.4 with $k = -1$, since $d(u, w) = \|w - u\|$ and $d(w, u) = \|u - w\| = \|(-1)(w - u)\|$. Axiom D.5 comes, as we would expect, from N.5. We have to show $d(a, c) + d(c, b) \geqslant d(a, b)$. Now $d(a, c) + d(c, b) = \|c - a\| + \|b - c\|$. If $u = c - a$ and $v = b - c$, then $u + v = b - a$. Substituting these values for u and v in axiom N.5 gives the desired result.

A vector space for which a norm, satisfying axioms N.1 to N.5, has been defined is called a *normed vector space*. As we have just proved, it is a metric space and so we can carry over to it the concepts and results established for metric spaces—the spheres $S(u, r)$, open balls $B(u, r)$, closed balls $\bar{B}(u, r)$, limits, Cauchy conditions for convergence.

A question concerned with visualization may be discussed here. You may be accustomed to visualizing a vector, u, as an arrow, and so find very odd the symbol $S(u, r)$ for a sphere with centre u. An arrow does not seem a very suitable thing for the centre of a sphere. However there is no real conflict. In plane geometry, (a, b) may represent the position of a point or the components of a vector. If we think of that vector as an arrow, then that arrow drawn from the origin fixes the position of the corresponding point. Conversely, given the origin, O, and a point, P, we can draw the arrow, OP, that connects them.

Incidentally, it may be remarked that in the picturesque language of functional analysis, every object, however complicated, is thought of as a point in some metric space. The distances between the points are significant but the points themselves do not in any way reflect the complexity of the objects they represent. In the same way a teacher discussing musical style might put two points, fairly close together on a blackboard, to represent Beethoven and Brahms, and a third point, somewhat removed from the other two, to represent Stravinsky. The distances, not the individual points, convey the message.

In section 3.1 we considered the iteration

$$\left.\begin{aligned}x_{n+1} &= 0.4x_n - 0.3y_n + 1\\ y_{n+1} &= 0.3x_n + 0.4y_n\end{aligned}\right\} \tag{1}$$

We now want to introduce a symbolic notation that will let us consider a certain type of iteration without being tied to this particular example. The equations (1) can be put in the form

$$w_{n+1} = u + Mw_n \tag{2}$$

with $w_n = (x_n, y_n)$, $u = (1, 0)$ and M denoting the function $(x, y) \rightarrow (0.4x - 0.3y, 0.3x + 0.4y)$. If in the generalized iteration form, (2), we take $w_0 = 0$, we find in turn $w_1 = u$, $w_2 = u + Mu$, $w_3 = u + Mu + M^2 u$ and generally $w_n = u + Mu + M^2 u + \ldots + M^{n-1} u$. A certain assumption has been used in making this calculation. Consider for example the stage where we find w_4 from w_3. We write $w_4 = u + Mw_3 = u + M(u + Mu + M^2 u) = u + Mu + M^2 u + M^3 u$. In the last step here we assume that the effect of M on the sum $u + Mu + M^2 u$ can be found by applying M to each of the terms and then adding the results. A similar assumption is made at the other stages of the iteration. We are in fact assuming that M is an *additive function*. That is to say, if r is any whole number, and $v_1, v_2, \ldots v_r$ are any r vectors, then

$$M(v_1 + v_2 + \ldots v_r) = Mv_1 + Mv_2 + \ldots + Mv_r. \tag{3}$$

Not every function is additive. For instance, among functions $\mathbb{R} \rightarrow \mathbb{R}$, the function 'squaring', $x \rightarrow x^2$, is not additive, though beginners at algebra seem to want to believe that it is.

Result 1.   If the iteration $w_{n+1} = u + Mw_n$ is defined in any vector space, if M is additive and if $w_0 = 0$, then for any whole number n, $w_n = u + Mu + M^2 u + \ldots + M^{n-1} u$.
This can be proved formally by induction.

Question.   We assumed $M0 = 0$. Does this follow from M being additive, or does it require a separate assumption?

We now want to consider the convergence of the iteration. In the particular iteration that we studied in section 3.1, the application of M halved the length of any vector. Clearly we can relax this condition. The essential point was that the lengths of the links of the chain formed a convergent geometrical progression. We can replace $\frac{1}{2}$ by any number k with $0 \leqslant k < 1$. There is a further way in which we can relax the conditions. It is not necessary that each link be exactly k times as long as the previous. It is as good, or better, if it is less than that. Accordingly we replace our earlier condition $\|Mv\| = 0.5\|v\|$ by the much less restrictive condition $\|Mv\| \leq k\|v\|$ for some k with $0 \leq k < 1$.

Result 2. For any normed vector space, $\mathscr{S}$, if M is an additive function, $\mathscr{S} \to \mathscr{S}$ for which $\|Mv\| \leq k\|v\|$ for every vector v, where $0 \leq k < 1$, then the series $u + Mu + M^2u + \ldots + M^nu + \ldots$ satisfies the Cauchy condition.

Proof. This proof uses essentially the same ideas as our argument for the special case in two dimensions, but perhaps it may be worth while to spell it out in detail. If s_k represents the sum of the first k terms of the series, we have to show that, for any $\varepsilon > 0$, $d(s_n, s_{n+p}) \leq \varepsilon$ for some n and for all the values $1, 2, 3 \ldots$ of p. By the definition of distance in a normed space $d(s_n, s_{n+p}) = \|s_{n+p} - s_n\| = \|M^nu + M^{n+1}u + \ldots + M^{n+p-1}u\|$. We now use the polygon inequality, which amounts to saying that the distance between two points of a chain cannot exceed the sum of the lengths of the links between them. It was indicated very early in section 3.1 that this principle was valid in any metric space. Accordingly, the last expression we had does not exceed $\|M^nu\| + \|M^{n+1}u\| + \ldots + \|M^{n+p-1}u\|$. We now use our information about M. Every time M acts, it produces a reduction in length, corresponding to multiplication by k or some smaller number. So, if $\|u\| = a$, $\|M^ru\| \leq k^ra$. Accordingly $\|M^nu\| + \|M^{n+1}u\| + \ldots + \|M^{n+p-1}u\| \leq k^na + k^{n+1}a + \ldots k^{n+p-1}a$.

However large p may be, the sum of the geometrical progression here will be less than its sum to infinity, which is $k^na/(1-k)$. This gives us an overestimate for the length of any part of the chain that occurs after the first n links. As $0 \leq k < 1$, the number $a/(1-k)$ is a fixed positive number (or 0 in the trivial case where $a = 0$), and by choosing n sufficiently large we can make the factor k^n as small as we wish. Thus we can find a stage after which the tail of the chain lies in a ball as small as anyone may demand. The series satisfies the Cauchy condition.

If the space $\mathscr{S}$ is *complete*, as it will be in the examples we shall consider, and M satisfies the conditions mentioned in results 1 and 2, the iteration $w_{n+1} = u + Mw_n$ will converge to a limit, whatever the vector u may be.

Question for investigation. If in this work we drop the condition $w_0 = 0$ and allow any vector for the initial vector w_0, can this cause the iteration to diverge?

This iteration would arise naturally in an attempt to solve the equation $w = u + Mw$ for w. It will do us no good if the iteration converges but converges to something that is not the solution of the equation. In fact, with the conditions stated, the iteration procedure will always yield a solution of the equation. The proof of this is deferred until section 5.7, so that we may proceed as soon as possible to examples of applications.

Note that it is not sufficient to put on M the condition that $\|Mv\| < \|v\|$ for every v. Such a condition would leave open the possibility that we might get a chain with links of lengths 1, $\frac{1}{2}$, $\frac{1}{3}$, $\frac{1}{4} \ldots$. The length of such a chain would be infinite, and would not place any restriction on our wanderings, even though each link is shorter than the previous one. Our approach requires that the ratio of the length of any link to that of the previous link must remain below some fixed fraction, which itself is less than 1. This is why we need the carefully phrased condition, $\|Mv\| \leqslant k\|v\|$, where $0 \leqslant k < 1$. M is then called a *contraction mapping* or a *contraction operator*.

3.3. The space of continuous functions

The space about to be described is the metric space of most importance for many numerical analysts. It is represented by the symbol $\mathscr{C}[a, b]$. Its elements are functions, $\mathbb{R} \to \mathbb{R}$, continuous on the closed interval $[a, b]$. It is a vector space, the basic vector operations being defined in the natural way. For two functions f and g, at any point x of the interval the function $f + g$ has the value $f(x) + g(x)$ and the function kf the value $kf(x)$.

It is not hard to verify that the vector axioms are satisfied. The zero vector, required by V.4 to be a vector the addition of which makes no difference, is the function $x \to 0$ for every x in the interval. Its graph lies in the x-axis. Subtraction, required by V.5, is a familiar operation. If $h = f - g$, then $h(x) = f(x) - g(x)$. So, for instance, if $f(x) \equiv x^2$ and $g(x) \equiv x^3$, then $f - g$ is the function $x \to x^2 - x^3$. The remaining axioms correspond to rules that we would use without a moment's thought if we were working with polynomials, trigonometric functions or any other continuous functions.

The norm now has to be defined. Figure 10 shows the graph of a function which in the interval $[a, b]$ rises to a height P and sinks to a depth Q, both measured away from the x-axis. The norm, $\|f\|$, is defined as the larger of these two numbers. This number is the

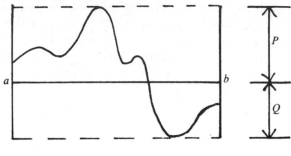

F<small>IG</small>. 10

maximum value of $|f(x)|$ for x satisfying $a \leqslant x \leqslant b$. In the literature you may meet this definition in the form $\|f\| = \max\{|f(x)| : x \in [a, b]\}$ or as $\|f\| = \sup\{|f(x)| : x \in [a, b]\}$.

We must of course check that this definition satisfies the norm axioms, N.1 to N.5. It is clear that $\|f\|$ is a real number (N.1) and that it is never negative (N.2). When $\|f\| = 0$, the height to which the graph rises and the depth to which it falls are both 0; that is to say, the graph lies in the x-axis. So $\|f\| = 0 \Rightarrow f = 0$, as required by N.3. Note that $f = 0$ means that f is the zero function, not that $f(x) = 0$ for some particular x. This is an example of a situation where the traditional terminology, 'the function $f(x)$' could be very confusing.

Axiom N.4 is easily verified. Axiom N.5 requires $\|f + g\| \leqslant \|f\| + \|g\|$. This again is almost obvious. Let $\|f\| = c$ and $\|g\| = d$. We want to show $\|f + g\| \leqslant c + d$. From the norms of f and g, we see that, for all x in the interval, $f(x)$ lies between $-c$ and c, and $g(x)$ lies between $-d$ and d. It follows that the sum, $f(x) + g(x)$ lies between $-c - d$ and $c + d$, as required. (The phrase 'between $-c$ and c' is to be interpreted here as permitting the values $-c$ and c themselves.)

Since $\mathscr{C}[a, b]$ is a normed space, it becomes a metric space when we introduce the definition of distance, $d(f, g) = \|f - g\|$, as was shown in section 3.2. Now $\|f - g\|$ is the maximum value of $|f(x) - g(x)|$ in the interval. It is now seen why the space $\mathscr{C}[a, b]$ is so important for numerical analysis. If f is some function we seek to compute and g is an approximation to it, a statement such as $\|f - g\| < 0.000\,001$ means that a table giving the approximate values, $g(x)$, will nowhere be as much as $0.000\,001$ away from the true values, $f(x)$. This is clearly a very useful type of statement to have.

Figure 11 illustrates the meaning of $\|f - g\|$. A vertical arrow is

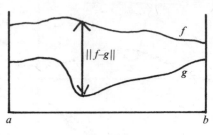

Fig. 11

drawn at the place where the graphs are farthest apart. The length of this arrow is $\|f - g\|$ and gives the distance between f and g in $\mathscr{C}[a, b]$.

Many problems in approximation can be expressed in terms of this distance. Weierstrass proved that any continuous function on $[a, b]$ can be approximated arbitrarily closely by a polynomial. Thus a polynomial can be found as close as we wish to any function f in $\mathscr{C}[a, b]$. We may ask what polynomial of degree 10, or less, is closest to a given f. In both of these sentences closeness is to be measured in terms of the distance as defined above.

This definition of distance shows what we are to understand by $\bar{B}(f, r)$, the closed ball with centre f and radius the number r. A function g will be in this ball if $\|g - f\| \leqslant r$. This means $|g(x) - f(x)| \leqslant r$ for each x in $[a, b]$, so $g(x)$ cannot be less than $f(x) - r$ or more than $f(x) + r$. In Figure 12, the region in which the graph of g must lie is bounded by the two broken lines. One of these is obtained by raising the graph of f through a distance, r, the other by lowering the graph of f the same distance. If the graph of g never goes outside this region, but meets its boundary (the broken lines) at one or more points, then $\|g - f\| = r$, and g lies on the sphere $S(f, r)$. If the graph of g lies between the broken lines, but never reaches either of them, then $\|g - f\| < r$ and g lies in the open ball, $B(f, r)$. If g satisfies either of these conditions, that is, if g is either in $S(f, r)$ or in $B(f, r)$, then g is in the closed ball, $\bar{B}(f, r)$.

Now suppose we have a sequence of continuous functions, $g_1, g_2, g_3, \ldots$ such that $g_n \to f$ in the metric space $\mathscr{C}[a, b]$. This means that, however small we may make the radius r, there will be a stage in the sequence after which all the g_n lie in $\bar{B}(f, r)$. This means that, however narrow we make the region between the broken lines

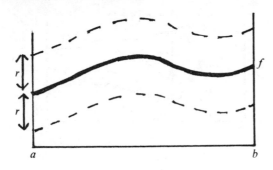

F<small>IG</small>. 12

in Figure 12, all the graphs of the g_n in some tail of the sequence will lie in that region.

In classical analysis, this would be described by saying that g_n *tends uniformly* to f. Before 1847 this concept was unknown (see Grattan Guinness, chapter 6). In earlier times $g_n \rightarrow f$ meant simply that $g_n(x)$ approached $f(x)$ for each value x. To-day this is called *pointwise convergence*. Pointwise convergence has a number of awkward features. One is that, even if all the functions g_n are continuous, their limit f need not be. For an example we need look no further than $g_n(x) \equiv x^n$ in $[0, 1]$. If $0 \leqslant x < 1$, $x^n \rightarrow 0$, but for $x = 1$, however large n may be, $x^n = 1$. Thus the graph of the limit function f consists of the x-axis for $0 \leqslant x < 1$ and the point $(1, 1)$. For this sequence, g_n does not approach f in the metric we have considered, for the distance $\|g_n - f\|$ does not approach 0. In Figure 13 the broken lines correspond to $r = 0.25$. However large n may be

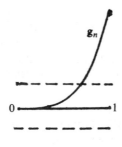

F<small>IG</small>. 13

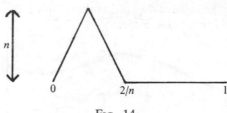

F<small>IG</small>. 14

chosen, we cannot make the graph of g_n stay between these lines. (Note that the definition of $\|g_n - f\|$ as the supremum of $|g_n(x) - f(x)|$ still makes sense, although f is not continuous, and therefore is not even in the space $\mathscr{C}[a, b]$.)

Another awkward thing about pointwise convergence is that we can have $g_n \to 0$ and at the same time $\int_0^1 g_n(x)\, dx$ does not tend to 0 and may even tend to infinity. Figure 14 shows a graph of g_n to illustrate this. Imagine the graph represents some membrane filled with an incompressible substance which ensures that the area of the triangle is always equal to 1. Then a heavy iron is brought in, and we push it to the left so that the triangle is made continually narrower. When the base of the triangle is $2/n$, its height must be n to keep the area equal to 1.

It is perhaps surprising, in view of the ever increasing height of the peak, that for every x in $[0, 1]$, $g_n(x) \to 0$. It is clear enough that $g_n(0) \to 0$, for in fact $g_n(0) = 0$ for all n. Now consider $x = c$ where $c > 0$. As the ironing process proceeds, a time will come after which the base of the triangle will have length less than c, and $g_n(c)$ will be 0. This in fact happens when $n > 2/c$. Hence $g_n(c) \to 0$ for any $c > 0$.

Now $\int_0^1 g_n(x)\, dx$ is the area under the graph, which is always 1. Thus

$$\int_0^1 g_n(x)\, dx \to 1$$

although $g_n(x) \to 0$ for each x. If we had decided to make the triangle of height n^2 instead of n, we could have made the integral tend to infinity.

You may feel this example is artificial because different lines are used to construct the graph. However it is perfectly possible to find analytic functions that show essentially the same behaviour, for example the function mentioned in the following exercise.

Exercise. Let $g_n(x) = (n^2 x)^n e^{-n^2 x}$. Make a table to show how g_n behaves in $[0, 1]$ for $n = 10$. Estimate

$$\int_0^1 g_{10}(x)\, dx.$$

Do the same two things for g_{20}.

Neither of the inconveniences discussed above can arise when $g_n \to f$ in the sense of uniform convergence, that is, convergence in the metric space $\mathscr{C}[a, b]$. It is a theorem in classical analysis that if all g_n are continuous, and g_n tends uniformly to f, then f also is continuous. It is also true that

$$\int_a^b g_n(x)\, dx \to \int_a^b f(x)\, dx.$$

This is very reasonable in the light of Figure 12. Raising the graph of f a distance r increases the area under the graph by $r(b-a)$; lowering through r reduces the area by the same amount. If g_n has a graph lying entirely between the dotted lines, its integral cannot differ from that of f by more than $r(b-a)$, and by taking a sufficiently late tail of the sequence, we can make r as small as we like.

Completeness of $\mathscr{C}[a, b]$. Finally we must consider the question, is $\mathscr{C}[a, b]$ complete? Does every sequence in $\mathscr{C}[a, b]$ that satisfies Cauchy's condition converge to a function in $\mathscr{C}[a, b]$?

We begin by picturing what it means that a sequence $g_1, g_2, g_3 \ldots$ satisfies Cauchy's condition. This means that, if the opponent produces a number ε_1, we can find a number s such that all the functions after g_s lie in $\bar{B}(g_s, \varepsilon_1)$. Similarly, to the opponents ε_2 there corresponds a closed ball $\bar{B}(g_t, \varepsilon_2)$ containing all the functions that follow g_t. The situation is shown in Figure 15. The region corresponding to $\bar{B}(g_s, \varepsilon_1)$ is between the broken lines; the region corresponding to $\bar{B}(g_t, \varepsilon_2)$ is shown shaded. Now consider the section of this diagram by the line $x = c$, corresponding to any c in $[a, b]$. If $n > s$, $g_n(c)$ must lie between the broken lines, hence in the interval HL. If $n > t$, $g_n(c)$ must lie in the shaded region, hence in the interval JK. In fact, we are going to find on this line a set of nested intervals, with lengths tending to nought. As this line is a representation of the real number system, $\mathbb{R}$, we have returned to

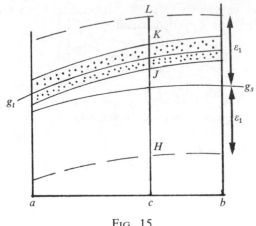

FIG. 15

the original setting of the Cauchy condition. We know $\mathbb{R}$ is complete, so we can be sure the intervals close down on a point, which
gives the limit of the sequence $g_n(c)$. Imagine this point marked, and
the same done for each vertical line that springs from the interval
$[a, b]$. These points define a function f, since they make a definite
value y correspond to each x in $[a, b]$.

We now have a construction for f. To prove what is required, we
have to show two things; (i) that the sequence g_n in $\mathscr{C}[a, b]$ tends to f,
(ii) that f is a continuous function, since otherwise it does not qualify
as a member of $\mathscr{C}[a, b]$.

It might seem that (i) is unnecessary, for our construction above
shows that $g_n(c)$ tends to a limit and defines $f(c)$ as that limit. But
this only shows that $g_n(x)$ tends to $f(x)$ for each x. It establishes
pointwise convergence, not the uniform convergence we need for a
limit in the metric of $\mathscr{C}[a, b]$.

However this gap is not too hard to fill. Suppose we have established that all the functions g_{N+p}, where $p = 1, 2, 3 \ldots$, make
$\|g_N - g_{N+p}\| \le \varepsilon$. Cauchy's condition assures us that such an N exists
for any $\varepsilon > 0$. So the graph of g_{N+p} lies between the broken lines in
Figure 16. It is clear from the construction, involving nested intervals, by which f was defined, that the graph of f also lies between
the broken lines. It is possible, as illustrated in Figure 16, that for
some x the values $f(x)$ and $g_{N+p}(x)$ may differ by as much as 2ε. But
since the vertical distance between the broken lines is 2ε, it is

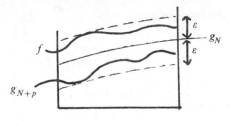

F IG . 16

impossible that the difference between $f(x)$ and $g_{N+p}(x)$ should be more than this. So, for all x in the interval, $|f(x) - g_{N+p}(x)| \leqslant 2\varepsilon$. This means $\|f - g_{N+p}\| \leqslant 2\varepsilon$. So there is a tail of the sequence lying in $\bar{B}(f, 2\varepsilon)$. The sequence does converge to f in the metric of $\mathscr{C}[a, b]$; it does converge uniformly to f.

A theorem, quoted earlier, states that if continuous functions, g_n, tend uniformly to f, then f must be continuous. This immediately establishes the second part of what we wished to prove.

Since convergence in $\mathscr{C}[a, b]$ is equivalent to uniform convergence, $\mathscr{C}[a, b]$ is sometimes described as the space of continuous functions with the uniform metric.

We have shown that $\mathscr{C}[a, b]$ is a vector space, that it is normed and that it is complete. Spaces with these properties were extensively studied by the great Polish mathematician, Stefan Banach. Spaces with all three properties are known as *Banach spaces*.

Exercises

1. Find the norm of the function f in $\mathscr{C}[0, 1]$ corresponding to each of the following expressions for $f(x)$;
(a) $3x + 4$.   (b) $x^2 - x$.   (c) $5x - 3$.
(d) $x^2 + x$.   (e) $\sin \pi x$.   (f) $-x^2 + x - 0.2$.
(g) $-x^2 + x - 0.1$.   (h) $(x^2 - x)^{10}$.

2. In $\mathscr{C}[0, 1]$ let p denote the function $x \to x$ and q the function $x \to 1 - x$. A function f is known to belong both to $\bar{B}(p, 0.5)$ and $\bar{B}(q, 0.5)$. Sketch the region in which the graph of f must lie. Can any function belong both to $B(p, 0.5)$ and $B(q, 0.5)$?

3. The functions f and g in $\mathscr{C}[0, 1]$ are defined by $f(x) \equiv a$ and $g(x) \equiv x^9(1 - x)^{11}$. For what value of a is the distance of f from g least? What is then the value of this distance? Can we reduce this distance by allowing f to be given by $f(x) = a + bx$ and choosing the most suitable values for a and b?

3.4. Iteration and integral equations

To solve the integral equation

$$f(x) = 1 + \int_1^2 \frac{1}{x+y} f(y)\, dy \tag{4}$$

it would be natural to try the iteration

$$g_{n+1}(x) = 1 + \int_1^2 \frac{1}{x+y} g_n(y)\, dy \tag{5}$$

If we take $g_0(x) \equiv 0$, this produces a sequence of continuous functions, $g_0, g_1, g_2 \ldots$. Will this sequence converge?

If we define the operator T by $Tg = h$, where

$$h(x) = \int_1^2 (x+y)^{-1} g(y)\, dy,$$

equation (5) may be written $g_{n+1} = 1 + Tg_n$. Here **1** stands for the function with the value 1 for each x in the interval. We shall only be interested in values of x lying in $[1, 2]$, so T is a mapping $\mathscr{C}[1, 2] \to \mathscr{C}[1, 2]$. T is additive, so iteration leads to the series $1 + T1 + T^2 1 + \ldots$. This will converge if T is a contraction operator. Can we find a useful estimate for the ratio of $\|Tg\|$ to $\|g\|$?

First of all, it is clear that, if on $[1, 2]$ g has positive values only, the same will be true of Tg. As the function **1** is positive, it follows that all the functions $T1, T^2 1, \ldots T^n 1, \ldots$ in the series take positive values only. Thus the norm of any one of these is given by its maximum value in the interval.

For x and y in $[1, 2]$, $(x+y)^{-1} \leqslant 0.5$. We shall certainly increase the value of $\int_1^2 (x+y)^{-1} g(y)\, dy$ if we replace $(x+y)^{-1}$ by its maximum value 0.5 and $g(y)$ by its maximum value $\|g\|$. The value of the amended integral is simply $0.5\|g\|$. So, for any x in the interval, $h(x) \leqslant 0.5\|g\|$, where, as before, h stands for Tg. So $\|Tg\|$, being the maximum value of $h(x)$, cannot exceed $0.5\|g\|$.

Accordingly $\|T^{n+1}1\| \leqslant 0.5\|T^n 1\|$. We are once again dealing with a chain where each link is at most half as long as the previous one. So the series converges and it is easy to estimate the magnitude of the terms; since $\|1\| = 1$, $\|T^n 1\| \leqslant 2^{-n}$. If we take only the terms up to and including $T^{20}1$, we neglect a length of chain that does not exceed the infinite sum $2^{-21} + 2^{-22} + 2^{-23} + \ldots$. This sum is 2^{-20}, slightly less than 0.000 001. Accordingly the table we obtain cannot anywhere differ from the true value by as much as 0.000 001.

There are two points to note. We have shown the series converges, but we have not proved that it converges to a solution of the equation (4). As in a previous instance, the proof of this is deferred until section 5.7. The second point is that the iteration soon leads to analytical complications. With $g_0(x) \equiv 0$, we get $g_1(x) = 1$, $g_2(x) = 1 + \ln(x+2) - \ln(x+1)$ and thereafter things become distinctly unpleasant. It would be necessary to use approximate numerical integration. Estimating the size of the errors this would introduce is a separate question, and will not be discussed here.

The method just used can be applied to a fairly wide class of problems. Let an operator T be defined by $Tg = h$ where

$$h(x) = \int_a^b K(x, y)g(y)\, dy,$$

for $a \leqslant x \leqslant b$. If the maximum value of $|K(x, y)|$ is M, it can be shown, by a very crude estimate, that $\|h\| \leqslant M(b-a)\|g\|$. If $M(b-a) < 1$, then T will be a contraction operator.

It is important to realize that, while T being a contraction operator is sufficient to ensure the convergence of the iteration, it is in no way necessary. Consider for example the integral equation

$$f(x) = 1 + \int_0^x f(y)\, dy, \tag{6}$$

where $0 \leqslant x \leqslant b$. Let $h = Tg$, with

$$h(x) = \int_0^x g(y)\, dy.$$

Here again we have the simplifying circumstance that we need only consider functions with positive values. Accordingly, $h(x)$ has its maximum value when $x = b$. The maximum height of the graph of g is $\|g\|$. It is clear from Figure 17 that $h(b)$, which equals the area under the graph of g for the interval $[0, b]$ cannot exceed $b\|g\|$. Accordingly T will be a contraction operator if $b < 1$.

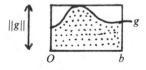

FIG. 17

However when we apply iteration to the integral equation (6), we obtain the series

$$1 + x + (x^2/2) + (x^3/6) + \ldots + x^n/(n!) + \ldots$$

This is the series for e^x and it is known to converge for all finite values of x. The contraction operator argument told us that, for positive values of x, it would converge in $[0, b]$ for $b < 1$. Evidently this approach can err very much on the side of caution. The reason for this is evident. The contraction operator principle is based solely on comparing the lengths of the links of a chain with a convergent geometrical progression. But there are many convergent series that behave in an entirely different way. For instance, the series for e^3 is $1 + 3 + 4.5 + 4.5 + 3.375 + 2.025 + 1.0125 + \ldots$. Initially the terms increase. Later, they decrease more rapidly than any geometrical progression.

If our earlier arguments about chains are examined, it will be found that all they really require is that the lengths of the links of the chain should form a convergent series. We can state this, perhaps rather loosely, as follows; if a chain has an infinite number of links but a finite total length, however it may be placed it will give a convergent sequence of points.

In more conventional terms we have the theorem; in any Banach space, the series of vectors, $\sum_1^\infty v_n$, must be convergent if the series of real numbers, $\sum_1^\infty \|v_n\|$, is convergent.

A theorem of this kind occurs already in classical analysis. If $\sum c_n$ is a series of real or complex numbers, it is known that $\sum c_n$ will be convergent if $\sum |c_n|$ is convergent. When this happens, the series c_n is said to be *absolutely convergent*, so the theorem just quoted takes the rather strange-sounding form; absolute convergence implies convergence.

A series can be convergent without being absolutely convergent. An example is the series $\ln 2 = 1 - (1/2) + (1/3) - (1/4) + (1/5) - \ldots$. When absolute values are taken, all the signs are positive and the sum is infinite. However such series have strange properties. Consider the following;

$$\ln 2 = 1 - (1/2) + (1/3) - (1/4) + (1/5) - (1/6) + (1/7) - (1/8) + \ldots$$
$$(1/2) \ln 2 = \quad (1/2) \quad\quad - (1/4) \quad\quad + (1/6) \quad\quad - (1/8) + \ldots$$

Add

$$(3/2) \ln 2 = 1 + (1/3) - (1/2) + (1/5) + (1/7) - (1/4) \ldots.$$

The series so obtained has the same terms as that for ln 2, but in a different order, yet its sum is different. In such series the sum depends on the order of the terms as well as on the numbers in the series.

What is happening may be visualized in the following way. Suppose we have two liquids which in some sense cancel each other out, like acid and alkali. We think of them as positive and negative. We have a container of positive liquid and a container of negative liquid, and we can let them flow in turn into a bowl.

For an absolutely convergent series there is a finite amount, p, of positive liquid, and a finite amount, n, of negative liquid. It does not matter in what manner we turn the taps on and off, provided that each part of the liquid comes out sometime; the liquid in the bowl is going to approach the value $p - n$.

It is quite otherwise with a series like that for ln 2. Now we have an infinite amount of each kind of liquid, and we can make the value of the liquid in the bowl approach any value we like. If in the long run the liquids emerge at the same rate, the limiting value will be 0. If we arrange for the positive liquid to emerge on the average twice as fast as the negative, the value will tend to $+\infty$. Similarly, by favouring the negative, we can approach $-\infty$. By suitable man-oeuvres we can get any value inbetween. We can also arrange to produce an oscillation that never settles down to any limit at all.

Exercise. Let $K(x, y)$ give a continuous function, defined for $0 \leqslant x \leqslant b$, $0 \leqslant y \leqslant b$, with $|K(x, y)| \leqslant M$. T is defined by $Tg = h$ where

$$h(x) = \int_0^x K(x, y)g(y)\, dy.$$

and $v_n = T^n v_0$ where $v_0 \in \mathscr{C}[0, b]$. Show that

$$v_1(x) \leqslant M\|v_0\|x, \quad v_2(x) \leqslant M^2\|v_0\|x^2/2$$

and generally that

$$v_n(x) \leqslant M^n\|v_0\|x^n/(n!).$$

Deduce that, however large b may be, the iteration $g_{n+1} = v_0 + Tg_n$ with $g_0 = 0$ is convergent. Does it make any difference if some other continuous function is chosen for the initial g_0?

3.5. A matrix iteration

It is not necessary to go to calculus to find an example of an iteration that converges and yet is not given by a contraction operator. Consider the iteration $v_{n+1} = Mv_n$, where $v_n = (x_n, y_n)$, a vector in the Euclidean plane, and M denotes the function

$$(x, y) \rightarrow (-3.5x + 10y, -1.2x + 3.5y)$$

or, in matrix notation

$$M = \begin{pmatrix} -3.5 & 10 \\ -1.2 & 3.5 \end{pmatrix}.$$

It is obvious that M is not a contraction operation. For instance $(0, 1) \rightarrow (10, 3.5)$, which represents an enlargement more than 10 times. Yet if $M^n v$ is tabulated for any vector v it will be found that $M^n v$ tends to 0 sufficiently rapidly for the series $v + Mv + M^2 v + \ldots$ to converge.

There is a standard procedure for seeing the meaning of a matrix transformation, namely, to find the eigenvalues and eigenvectors. The usual algorithm shows that M has eigenvector $(1, 0.3)$ with eigenvalue -0.5, and the eigenvector $(1, 0.4)$ with eigenvalue 0.5. This means that any point in the line $y = 0.3x$ goes to the point of that line, on the opposite side of the origin, that is half as far from the origin as the original point. Any point in the line, $y = 0.4x$, stays in that line, on the same side of the origin and at half the original distance. The transformation can be put in its simplest form by taking these two invariant lines as axes of co-ordinates. The new co-ordinates (X, Y) can be introduced by writing

$$\begin{pmatrix} x \\ y \end{pmatrix} = X \begin{pmatrix} 1 \\ 0.3 \end{pmatrix} + Y \begin{pmatrix} 1 \\ 0.4 \end{pmatrix}.$$

Solving for X and Y gives $X = 4x - 10y$, $Y = -3x + 10y$.

If we choose for v_0 the vector $(0, 1)$, which we noticed was considerably enlarged when M acted on it, we obtain the following table for $M^n v$. In this table we record the vectors $M^n v$, both as they appear in the original (x, y) system of co-ordinates and as they appear in the (X, Y) system.

The table shows the clarity that comes from basing co-ordinates on invariant lines. In the Y-column each entry is half the entry above it; the same, accompanied by a change of sign, holds for the

n	x	y	X	Y
0	0	1	−10	10
1	10	3.5	5	5
2	0	0.25	−2.5	2.5
3	2.5	0.875	1.25	1.25
4	0	0.0625	−0.625	0.625
5	0.625	0.2188	0.3125	0.3125
6	0	0.156	−0.1563	0.1563
7	0.1563	0.0547	0.0781	0.0781

X-column. The convergence, and the rate of convergence, are clearly apparent.

In the x- and y-columns a fluctuation is seen, the magnitude of the numbers alternately increasing and decreasing. The vector (x, y) is given by $X(1, 0.3) + Y(1, 0.4)$ as we had earlier. We can see how (x, y) manages to increase in length even though X and Y steadily decrease in magnitude. The angle between the eigenvectors, $(1, 0.3)$ and $(1, 0.4)$ is very small, about $5°$. When X and Y have opposite signs, as they have for v_0, they are pulling in almost opposite directions. At the next stage, for v_1, their contributions are nearly in line and reinforce each other.

In this particular example, a certain regularity is noticeable even in the x, y columns. The x, y entries in any row are one-quarter of the entries two rows higher. This corresponds to the fact that

$$\begin{pmatrix} -3.5 & 10 \\ -1.2 & 3.5 \end{pmatrix}\begin{pmatrix} -3.5 & 10 \\ -1.2 & 3.5 \end{pmatrix} = \begin{pmatrix} 0.25 & 0 \\ 0 & 0.25 \end{pmatrix}.$$

Thus, although M is not a contraction operator, M^2 is. It multiplies all lengths by $\frac{1}{4}$. This gives us an alternative way of showing that the series $\sum M^n v$ converges. For we have

$$v + Mv + M^2v + M^3v + M^4v + M^5v + \ldots$$
$$= (v + Mv) + M^2(v + Mv) + M^4(v + Mv) + \ldots$$
$$= u + Nu + N^2u + \ldots,$$

where $u = v + Mv$ and $N = M^2$. As N is a contraction operator, the series must converge.

A similar argument shows that the series $\sum T^n v$ converges if some power of T, say T^k, is a contraction operator.

The argument involving eigenvalues and eigenvectors has wide application. It can be shown that the series $\sum T^n v$ is convergent for every vector v if T is a square matrix (with any number of rows) such that $|\lambda| < 1$ for each of its eigenvalues, λ. A proof will be found in Berezin and Zhidkov, volume 2, pages 46–47.

4 Minkowski spaces

4.1. Introduction

PYTHAGORAS' THEOREM in Euclidean geometry ascribes the length s to the vector (x, y) where $s^2 = x^2 + y^2$. In the theory of curved surfaces and in general relativity more complicated formulas occur, in which ds^2 is given by a quadratic differential expression. It is very noticeable that we always seem to be dealing with squares, either s^2 or ds^2. A student might wonder whether an obsession with squares has created a kind of mental prison for us. Is there any reason why a geometry must be based on squares? Could we have a geometry in which $s^p = x^p + y^p$, or in higher dimensions

$$s^p = \sum_1^n x_r^p,$$

where p is some number other than 2?

This idea was developed by Minkowski in his book, *Geometrie der Zahlen*, published in 1896. In this book and in earlier papers, he obtained interesting results in the theory of numbers by simple geometrical arguments. He defined the length, s of the vector $(x_1, \ldots, x_n)$ by

$$s^p = \sum_1^n |x_r|^p.$$

Absolute values have to be brought in to prevent vectors having zero or negative lengths. For instance, if $p = 3$, $s^3 = p^3 + q^3$ would make $(1, -1)$ of length (0) and $(0, -1)$ of length -1, which is contrary to our ideas of how distances should behave.

The systematic study of metric spaces dates from Frechét's thesis in 1906. Minkowski's definition of distance provided an example of a normed vector space some years before this general idea had been recognized. The space with

$$\|(x_1, \ldots, x_n)\| = \left[\sum_1^n |x_r|^p \right]^{1/p}$$

is now referred to as ℓ_p. The same symbol is used—and this indeed is its commonest use—for an infinite dimensional space with a similar norm. In chapter 2, mention was made of ℓ_p spaces in connection with the particular cases, $p = 1$, 2 and ∞.

In numerical analysis the ℓ_p spaces, other than ℓ_1, ℓ_2 and ℓ_∞, are used relatively rarely. Since this book is oriented towards applications, an explanation is in order as to why this chapter, dealing with the general ℓ_p space, is included.

There are two reasons. One is that many books, which a student may want to read or be required to read for examination purposes, introduce ℓ_p spaces very early, and start with a proof that seems designed to put students off the subject for life. An explanation of the ideas behind this proof will be given in section 4.2 in the hope that it will remove a very considerable obstacle to further study.

A second reason is that this explanation gives a good illustration of the way in which geometrical thinking helps work in functional analysis, and it leads on to concepts which are of importance for numerical analysis.

One naturally wonders—how does it come about that such frustrating proofs as that shortly to be considered get into the textbooks? The reason lies in the differences between discovering and publishing. When you are making a discovery you are guided by certain ideas, perhaps pictorial, which are helpful but very hard to pin down in a precise manner. When a mathematician is preparing a paper for publication, his or her first concern is to convince other experts that the theorem is really true, that it has been proved beyond all doubt. The first consideration, then, is logic and the proof eventually published may be entirely different from the train of thought that led the author to the discovery. Certainly a logical proof is essential for mathematics, both in learned journals and in expository textbooks. The trouble comes when eminent mathematicians, having concentrated so long on logical perfection, give the formal proofs without any indication of how these are arrived at. The psychological processes, involving guesswork and imagination, that lead to discoveries are quite as important as logic for students who wish to understand mathematics and perhaps to make their own mathematical discoveries.

4.2. Analysis of an obscure proof

The result to be proved is simple and natural enough. In a normed vector space, the triangle inequality is expressed by the axiom N.5, $\|u\| + \|v\| \leqslant \|u + v\|$. We want to show that this axiom is satisfied with Minkowski's definition of distance, given by

$$\|(x_1, \ldots x_n)\| = \left[\sum_1^n |x_r|^p\right]^{1/p} \quad \text{where} \quad p > 1.$$

A brief discussion shows that it is sufficient to prove the inequality in question for vectors that do not have any negative components. That is, we want to prove for $x_r \geq 0$, $y_r \geq 0$, the inequality

$$\left(\sum x_r^p\right)^{1/p} + \left(\sum y_r^p\right)^{1/p} \leq \left[\sum (x_r + y_r)^p\right]^{1/p} \quad (1)$$

A proof often given is sketched below. The remarks in italics represent questions that might occur to and puzzle even a very good and well-informed student, meeting this proof for the first time.

The proof begins by establishing a preliminary result, Hölder's inequality. If $u_r \geq 0$, $v_r \geq 0$ for $1 \leq r \leq n$, then

$$\sum_1^n u_r v_r \leq \left(\sum_1^n u_r^p\right)^{1/p} \left(\sum_1^n v_r^q\right)^{1/q}$$

where $(1/p) + (1/q) = 1$. (*Whatever is this result? However did anyone think of it? It does not look much like what we are trying to prove; what suggests that it might help? And whatever is this number q doing here?*)

We need not consider how Hölder's inequality is proved. Next in the proof, equation (2) is written.

$$\sum_1^n (x_r + y_r)^p = \sum_1^n x_r (x_r + y_r)^{p-1} + \sum_1^n y_r (x_r + y_r)^{p-1} \quad (2)$$

(*This is obviously true—but why do we split it this way?*)
Next apply Hölder's inequality to both parts.

(i) Let $u_r = x_r$, $v_r = (x_r + y_r)^{p-1}$. This gives

$$\sum x_r (x_r + y_r)^{p-1} \leq \left[\sum x_r^p\right]^{1/p} \left[\sum (x_r + y_r)^{(p-1)q}\right]^{1/q}.$$

As $(1/p) + (1/q) = 1$, it follows that $(p-1)q = p$ so this inequality can be rewritten as

$$\sum x_r (x_r + y_r)^{p-1} \leq \left[\sum x_r^p\right]^{1/p} \left[\sum (x_r + y_r)^p\right]^{1/q}.$$

(ii) A similar calculation, applied to the second part of the right-hand side of inequality (2), gives

$$\sum y_r(x_r + y_r)^{p-1} \leq \left[\sum y_r^p\right]^{1/p}\left[\sum (x_r + y_r)^p\right]^{1/q}.$$

The last two inequalities are added. On the left-hand side we find $\sum (x_r + y_r)^p$, which we will call S for short. On the right-hand side there is a common factor; it is $S^{1/q}$. Dividing by this common factor, we get $S^{1-(1/q)}$ on the left hand-side. As $1-(1/q)=1/p$, this left-hand side is now $S^{1/p}$. Replacing S by the expression it represents, we have now arrived at

$$\left[\sum (x_r + y_r)^p\right]^{1/p} \leq \left(\sum x_r^p\right)^{1/p} + \left(\sum y_r^p\right)^{1/p}$$

which is exactly what we wanted to prove. (*How on earth was this foreseen?*)

This concludes the summary of the proof as it is often met. How can we discover the ideas that lead to such a proof? As the ℓ_p space is a generalization of Euclidean space, a good starting point is to look at the problem in that familiar setting. Even in the Euclidean case, ℓ_2, the problem is not too simple in algebraic terms. We have to show

$$\sqrt{\left(\sum x_r^2\right)} + \sqrt{\left(\sum y_r^2\right)} \geq \sqrt{\left[\sum (x_r + y_r)^2\right]}.$$

However there is a very simple geometric proof. In Figure 18, the vectors OA, AB and OB represent x, y and $x + y$. We want to show $\|OA\| + \|AB\| \geq \|OB\|$. If we draw $AM \perp OB$, we have $\|OA\| \geq \|OM\|$ and $\|AB\| \geq \|MB\|$, since no side of a right-angled triangle can be longer than the hypotenuse. As $\|OB\| = \|OM\| + \|MB\|$, we only have

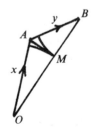

FIG. 18

to add the two inequalities to get the desired results. We need not bother about the possibility of M falling outside OB, as this cannot happen with $x_r \geqslant 0$, $y_r \geqslant 0$, which we are still assuming.

This construction already gives us a hint as to why, in the proof considered earlier, an expression is split into two parts and an inequality written for each. This corresponds to breaking OB into OM and MB. It also shows why the cancelling of a factor is to be expected in the algebra. When $p = 2$, the left-hand side of inequality (2) represents $\|OB\|^2$, which is equal to $\|OB\| (\|OM\| + \|MB\|)$, which is then shown not to exceed $\|OB\| (\|OA\| + \|AB\|)$. The factor $\|OB\|$ has to be removed to give the triangle inequality.

When we seek to generalize this argument from Euclidean, ℓ_2 space to Minkowski's ℓ_p space, we lose the possibility of constructing M by dropping a perpendicular from A to OB, since 'perpendicular' is no longer defined. However this obstacle is easily overcome. The circles through M, with centres O and B, touch at M and AM is their common tangent, as is indicated in Figure 18. The required inequalities, $\|OA\| \geqslant \|OM\|$ and $\|AB\| \geqslant \|MB\|$, are both instances of the fact that no point of a tangent to a circle can be nearer to the centre than the point of contact.

All of this generalizes without difficulty to the space ℓ_p. We now seek a point M on OB, such that AM is the tangent to the sphere, $S(O, r)$, that passes through M. It then turns out that AM is also tangent to the sphere $S(B, R)$ that passes through M.

The sign $\| \ \|_p$ is used to indicate the norm for the space ℓ_p. Our aim is to prove

$$\|x\|_p + \|y\|_p \geqslant \|x + y\|_p,$$

and our plan is to do this by showing

$$\|x\|_p \geqslant \|OM\|_p \quad \text{and} \quad \|y\|_p \geqslant \|MB\|_p.$$

Both of these we expect to demonstrate with the help of a general inequality, expressing that no point on the tangent to an ℓ_p sphere is closer (in the ℓ_p metric) to the centre of the sphere than the point of contact. Chapter 9 of this book is concerned with inequalities. In section 9.2 it is shown that someone setting out to express this geometrical fact would quite naturally arrive at Hölder's inequality. That is why Hölder's inequality comes into the proof at the beginning of this section.

One might guess that Hölder arrived at his rather strange inequality in this way, but in fact he discovered it in quite a different manner. In 1887 L. J. Rogers had published an inequality in *The Messenger of Mathematics*. Hölder observed expressions in it that looked like formulas for centres of gravity. He extracted a general procedure for obtaining inequalities. If a number of masses are placed on a curve, $y = f(x)$ that bends upwards (that is, has $f''(x) > 0$ everywhere), their centre of gravity will lie above the curve. The algebraic expression of this is an inequality. Hölder tried various function for f, and obtained the corresponding inequalities. Most of these were already known. However among his results was a new one—Hölder's inequality, obtained by taking $f(x) = x^p$ with $p > 1$ and then making certain substitutions.

It may be mentioned that if $a = (a_1, \ldots, a_n)$ and $x = (x_1, \ldots, x_n)$, Hölder's inequality can be written in the concise (and significant) form $|\sum a_r x_r| \leq \|x\|_p \|a\|_q$, where, as before $(1/p) + (1/q) = 1$. This inequality thus ties together two different Minkowski spaces, ℓ_p and ℓ_q.

4.3. A general construction for a norm

It would be possible to follow out the ideas of the last section, and show in detail how this geometrical approach, translated into algebra, enables us to construct the proof described earlier. This however will not be done, but another line will be followed, suggested by Figure 18. That figure is extremely simple, and the proof it suggests seems to depend on two properties only of the circle, (i) than no tangent ever enters the open ball bounded by the circle, (ii) that the point of contact of the circles with centres O and B is at M, a point on the line segment joining O and B. If this were not so we would be unable to use the equation $\|OB\| = \|OM\| + \|MB\|$ in our proof. Now these two properties are by no means confined to circles; many other curves possess them. This suggests that it may be fruitful to enquire what properties must be possessed by a family of curves, if these curves are to play the role of the spheres, $S(v, r)$, in some normed space. We will look at the norm axioms in turn, and see what they require of these curves.

Axiom N.4 states $\|kv\| = |k| \|v\|$. This tells us two things. First of all, if $k > 0$, it tells us that $S(O, k)$ is obtained from $S(O, 1)$ by a change of scale, in fact by the dilation $v \rightarrow kv$. Second, if we take $k = -1$, it shows that $\|-v\| = \|v\|$. So, if v is a point of $S(O, 1)$, the point $-v$ also must be on $S(O, 1)$. In other words, $S(O, 1)$ must be symmetric about the origin.

In a normed space, the distance $d(u, v)$ is defined as $\|u - v\|$. This means that a sphere with centre c can be obtained from a sphere with centre O, rather as if we had the spheres drawn on a transparent sheet that could slide (without rotation) over the plane. For suppose u is a point on the sphere $S(O, r)$. This means $\|u\| = r$. Then $u + c$ will be on the sphere $S(c, r)$, for $d(u + c, c) = \|(u + c) - c\| = \|u\| = r$.

The fact that tangents keep outside a circle is due to the fact that the closed ball, (the circle with the points inside it), is convex. A region is called convex if the line segment joining any two points of the region lies entirely in the region. If a moving point starts at position u at time $t = 0$ and travels with constant speed in a straight line to reach v at time $t = 1$, its position at time t is given by $u + t(v - u)$, which equals $(1 - t)u + tv$. If a region is convex, and the points u and v belong to it, so must every point, $(1 - t)u + tv$, with $0 \leqslant t \leqslant 1$. This is sometimes expressed more symmetrically; if a region is convex and contains the points u and v, it must contain every point $au + bv$ for which $a \geqslant 0$, $b \geqslant 0$ and $a + b = 1$.

In Figure 18, the convexity of the balls led us to a plausible argument for the validity of axiom N.5. It works the other way too; we can prove that, in any normed space, $\bar{B}(O, 1)$ is convex. It follows of course from this that every $\bar{B}(c, r)$ is convex.

To prove this, let u and v be points of $\bar{B}(O, 1)$. We want to prove that $au + by$ also is in the ball, provided $a \geqslant 0$, $b \geqslant 0$ and $a + b = 1$.

Now $\|au + bv\| \leqslant \|au\| + \|bv\|$ by N.5,

$\qquad\qquad = a\|u\| + b\|v\|$ since $a \geqslant 0$, $b \geqslant 0$,

$\qquad\qquad \leqslant a + b$ since $\|u\| \leqslant 1$, $\|v\| \leqslant 1$.

$\qquad\qquad = 1$ by hypothesis.

Hence $au + bv \in \bar{B}(O, 1)$.   Q.E.D.

Any ray through the origin must meet $S(O, 1)$ at a point distinct from the origin. For let w be any point, other than O, on this ray. By N.3, $\|w\| \neq 0$; so we can let $\|w\| = 1/k$. Then $\|kw\| = k\|w\| = 1$, so we have found a vector of norm 1 on the ray. What is being shown here is that O must be an interior point of $\bar{B}(O, 1)$.

The interesting thing is that these arguments are reversible. If we have a curve, symmetric about the origin, surrounding the origin,

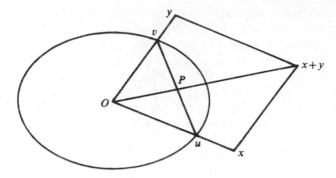

F<small>IG</small>. 19

and forming the boundary of a convex region, then that curve can play the role of $S(O, 1)$ in some normed space. Since $S(O, 1)$, so obtained, is symmetrical, by taking $S(O, k)$ as the curve obtained from $S(O, 1)$ by the dilation $v \rightarrow kv$, we can ensure that axiom N.4 is satisfied.

There are only five norm axioms, and of these N.5 is the only one that gives us cause for thought. It is not too obvious how to prove it. We want to show that $\|x + y\| \leqslant \|x\| + \|y\|$ for any two vectors, x, y, and it is not clear how we are to relate these two vectors to the properties of the unit ball. Figure 19 shows a way of achieving this. Here u and v are points on $S(O, 1)$, lying on the rays joining the origin to x and y. As u and v are on $S(O, 1)$, it follows that $\|u\| = 1$ and $\|v\| = 1$. Let $x = hu$ and $y = kv$. We have arranged things so that axiom N.4 holds, so we can deduce that $\|x\| = h \|u\| = h$, and $\|y\| = k \|v\| = k$. Our task is to prove $\|x + y\| \leqslant h + k$. As the convexity of the unit ball is part of our data, presumably we have to consider points on the line segment joining u and v. Since $x + y$ is to be involved, it will probably be best to consider the point, P, of that segment that lies on the ray from the origin to the point $x + y$. Since P is on the ray going to $x + y$, we must have $P = t(x + y)$ for some t. Hence $P = thu + tkv$. To be on the line segment, P must be expressible as $au + bv$ with $a + b = 1$. So $a = th$, $b = tk$ and $th + tk = 1$. Accordingly $t = 1/(h + k)$, and so $P = t(x + y) = (x + y)/(h + k)$. By the convexity property, P must lie in the unit ball, so $\|(x + y)/(h + k)\| = \|P\| \leqslant 1$. By N.4, this means $\|x + y\| \leqslant h + k$, that is to say, $\|x + y\| \leqslant \|x\| + \|y\|$, and N.5 is proved.

For convenience of exposition, this topic has been discussed in

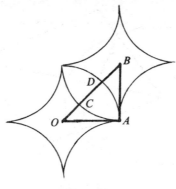

Fig. 20

terms of two-dimensional space, but this has nowhere affected the substance of the argument. What we have reached is a geometrical description of the unit ball in any Banach space—a convex region, symmetrical about the origin, with the origin as an interior point.

Convexity explains why we never speak of ℓ_p spaces with $0 < p < 1$. The two star-shaped curves in Figure 20 show what the spheres would look like if such values of p were allowed. In this figure the spheres $S(O, 1)$ and $S(B, 1)$ are shown. B is the point $(1, 1)$. As A and C lie on $S(O, 1)$, the norms $\|OA\|$ and $\|OC\|$ are both 1. Similarly from consideration of $S(B, 1)$ we see $\|DB\| = \|AB\| = 1$. The triangle inequality no longer holds, for $\|OA\| + \|AB\| = 2$, while $\|OB\| = \|OC\| + \|CD\| + \|DB\| = \|CD\| + 2$. The 'distance' from O to B is more if we travel in a straight line than if we go via A.

In chapter 2 we met the unit sphere for ℓ_1, the tilted square. As p grows upwards from 1, the unit ball is steadily inflated. It has reached the form of a circle when $p = 2$. By $p = 10$, it looks very much like a square with rounded corners, and for $p = 100$ it is indistinguishable (by eye) from a square with sides parallel to the axes of co-ordinates. We met this square as $S(O, 1)$ for the norm $\|(x, y)\| = \max\{|x|, |y|\}$, which, for this reason, receives the label ℓ_∞.

5 Linear operators and their norms

5.1. Linear operators

AT THE BEGINNING of calculus, students who have learnt how to differentiate x^n find it very easy to differentiate any polynomial. Any term ax^n gives $a(nx^{n-1})$ and then these new expressions have to be linked by + and − signs corresponding to those in the original polynomial.

The formal justification for this procedure lies in the two properties of differentiation, $D(u+v) = Du + Dv$, and $D(ku) = kDu$, where D represents the operation of differentiation, and k represents any constant. These two properties express the fact that D is a linear operator. We are now going to seek a general definition of this concept.

The equations above involve the expressions $u+v$ and ku, associated with the basic operations of pure vector theory. Thus our general definition must assume the inputs belong to some vector space. The outputs, Du and Dv, also appear added, and Du is multiplied by k. Thus the outputs also must lie in a vector space. Accordingly our definition runs; if L is a function $\mathcal{X} \to \mathcal{Y}$, where $\mathcal{X}$ and $\mathcal{Y}$ are vector spaces, L is called a linear operator if, for any vectors u, v in $\mathcal{X}$ and for any real number k, we have $L(u+v) = Lu + Lv$ and $L(ku) = kLu$.

The space $\mathcal{Y}$ may coincide with the space $\mathcal{X}$ but it does not have to. For instance, if f is a continuous function defined on $[0, 1]$ and

$$g(x) = \int_0^x f(t)\, dt,$$

then $f \to g$ is a linear function $\mathscr{C}[0, 1] \to \mathscr{C}[0, 1]$. On the other hand, if $s = \int_0^1 f(t)\, dt$, $f \to s$ is a linear function, $\mathscr{C}[0, 1] \to \mathbb{R}$.

Exercises on linear functions. In the following situations, say whether the function specified is linear or not.

1. Projection, $\mathbb{R}^3 \to \mathbb{R}$; $(x, y, z) \to x$.

2. $\mathbb{R} \to \mathbb{R}^3$, $x \to (x, x, x)$.

3. $\mathbb{R} \to \mathbb{R}$, $x \to x + 1$.

4. Rotation, $\mathbb{R}^2 \to \mathbb{R}^2$. $(x, y) \to (-y, x)$.

5. $\mathscr{C}[0, 1] \to \mathbb{R}$. $f \to f(0)$.

6. $\mathbb{R}^2 \to \mathbb{R}$, $(x, y) \to \sqrt{(x^2 + y^2)}$.

7. $\mathscr{C}[0, 1] \to \mathbb{R}$, $f \to \|f\|$.

8. $\mathbb{R}^3 \to \mathbb{R}$, $(x, y, z) \to \|(x, y, z)\|_\infty$.

9. $\mathbb{R}^3 \to \mathbb{R}$, $(x, y, z) \to \|(x, y, z)\|_1$.

10. $\mathbb{R}^3 \to \mathbb{R}$, $(x, y, z) \to x + y + z$.

11. $\mathscr{C}[0, 1] \to \mathscr{C}[0, 1]$, $f \to g$, $g(x) = [f(x)]^2$.

12. $\mathscr{C}[0, 1] \to \mathscr{C}[0, 1]$, $f \to g$, $g(x) = f(x^2)$.

13. $\mathscr{C}[0, 1] \to \mathbb{R}^6$, $f \to v$, where v is the vector $[f(0), f(0.2), f(0.4), f(0.6), f(0.8), f(1)]$. (This correspondence is involved when we deal with a function specified by a table.)

14.

$$\mathscr{C}[0, 1] \to \mathbb{R}. \quad f \to \int_0^1 f(x)\, dx - 0.5[f(0) + f(1)].$$

(This has to do with the error when the area under a curve is estimated by the trapezium rule.)

15. Difference operator. $f \to g$ where $g(x) = f(x + h) - f(x)$.

Much classical analysis was concerned with linear differential and integral equations, since these were much easier to deal with than non-linear ones. In algebra much attention was paid to linear transformations. The equations,

$$y_i = \sum_j a_{ij} x_j,$$

with $1 \le i \le m$ and $1 \le j \le n$, define a linear transformation from the space $\mathscr{X}$, consisting of vectors $(x_1, \ldots x_n)$, to the space $\mathscr{Y}$, consisting of vectors $(y_1, \ldots y_m)$. In matrix notation these equations appear as $y = Ax$.

The difficulty of dealing with non-linear problems continually steered classical mathematics in the direction of work on linear problems. The great power of modern computing facilities has changed this, and today the most important computational applications of functional analysis are to non-linear problems. This however does not mean that linear operators have ceased to be important. As we have just observed, differentiation and integration are linear operators; this however in no way implies that they can only be used with straight-line graphs. Perhaps the most striking experience of a young mathematics student is the sense of immense power felt after learning the elements of calculus and seeing the variety of their applications. In functional analysis also there are linear operators, generalizing differentiation and integration, and having a wide field of applications.

Matrices provide an elementary example of linear operators and can help us by suggesting how linear operators in general are likely to behave. Two matrices, each having m rows and n columns, can be added. If A, B, C are matrices, we understand by the equation $C = A + B$ that $Cv = Av + Bv$ for every vector v in the input space. We have essentially the same definition of sum for linear operators. If A and B are linear operators, each mapping a vector space $\mathcal{X}$ to a vector space $\mathcal{Y}$, then $A + B$ is defined as a function $\mathcal{X} \rightarrow \mathcal{Y}$, such that $(A + B)x = Ax + Bx$ for every x in $\mathcal{X}$.

If P and Q are matrices having a product PQ, then $w = PQu$ is equivalent to $w = Pv$ and $v = Qu$. If we imagine P and Q embodied in computer programs, the output of Q must be acceptable as an input for P. More formally, if Q is a function $\mathcal{X} \rightarrow \mathcal{Y}$ and P is a function $\mathcal{Y} \rightarrow \mathcal{Z}$, then PQ is a function $\mathcal{X} \rightarrow \mathcal{Z}$, and it sends x to z, where $z = Py$ and $y = Qx$.

We can also define kT for any linear operator, T, and any number k. We simply make $(kT)v = k(Tv)$.

The considerations above show that we can add linear operators that map $\mathcal{X}$ to $\mathcal{Y}$, and also multiply them by numbers—in short, we can define the two basic operations of pure vector theory. It is not hard to verify that such addition and multiplication comply with the requirements of the vector axioms, V.1 to V.10. Accordingly these linear operators form a vector space. This may at first seem a little confusing. When we work with vectors and matrices we have to be very careful to distinguish between them; the vectors lie in some space and the matrices define operations, such as dilation, rotation, reflection, shearing and so forth, applied to this space. Indeed this distinction is important and has to be maintained. But it is not really so surprising that these operations themselves form a vector space. The general 2×2 matrix satisfies the equation

$$\begin{pmatrix} a & b \\ c & d \end{pmatrix} = a\begin{pmatrix} 1 & 0 \\ 0 & 0 \end{pmatrix} + b\begin{pmatrix} 0 & 1 \\ 0 & 0 \end{pmatrix} + c\begin{pmatrix} 0 & 0 \\ 1 & 0 \end{pmatrix} + d\begin{pmatrix} 0 & 0 \\ 0 & 1 \end{pmatrix}.$$

Thus it is an arbitrary linear combination of the four ingredients shown above. This corresponds exactly to our childish picture of a vector space based on four animals, only here the animals are 2×2 matrices.

In human communication, it is certainly convenient to write

$$\begin{pmatrix} a & b \\ c & d \end{pmatrix}$$

with the letters a, b, c, d in the positions they have in the equations this matrix symbolizes,

$$y_1 = ax_1 + bx_2$$
$$y_2 = cx_1 + dx_2$$

However, when we instruct a computer to deal with this matrix, we cannot use such an arrangement of rows·and columns. We can only feed in the plain sequence of numbers, (a, b, c, d), which looks like, and indeed is, a vector in four dimensions.

In the same way, $n \times n$ matrices form a vector space of n^2 dimensions, and $m \times n$ matrices a vector space of mn dimensions.

If A and B are $n \times n$ matrices, their product is also an $n \times n$ matrix. In most vector spaces no operation of multiplying two vectors is defined. However the fact that such an operation exists does not disqualify a system from being a vector space. The qualifications for a vector space are positive, not negative. If a system passes all the tests provided by the vector axioms, V.1 to V.10, it is a vector space. The fact that it has other properties as well is irrelevant.

The matrix representation of linear operators that act on finite-dimensional spaces has been used to illustrate the fact that linear operators, $\mathcal{X} \to \mathcal{Y}$, form a vector space, but of course this representation is not necessary for the proof of the general result. That proof consists in showing that the vector properties, V.1 to V.10, are a logical consequence of the definition of linear operator and the definitions of $A + B$ and kA given above. We shall not go into the details of this proof.

5.2. Operator norms

In computing, as has been mentioned already, we are rarely dealing with exact values. If we compute $w = Av$, where v is a vector and A is a linear operator, we are naturally interested in knowing how large an error in the output may arise from the possible error in the input. If v_0 and w_0 are the true values, $w_0 = Av_0$, so the error $w - w_0 = Av - Av_0 = A(v - v_0)$. Thus if $e = v - v_0$ is the input error, the output error $w - w_0 = Ae$. This is characteristic of linear operators; the errors are transformed by exactly the same formula as the vectors themselves. Thus it does not matter whether we enquire how much errors can be enlarged, or how much vectors themselves can be enlarged. The former question is the most natural one in a computing context; the latter is more usual in a pure mathematical context.

In speaking of enlargement we are assuming that there are norms, both in the input and the output spaces, to measure how large a vector is. The enlargement of error is measured by the ratio $\|Ae\|/\|e\|$. We want to define the norm of A in such a way that it gives us a ceiling for this number. The formal definition is thus

$$\|A\| = \sup\{\|Ae\|/\|e\| : e \in \mathcal{X}, \quad e \neq 0\}.$$

The stipulation $e \neq 0$ has to be included, because the ratio is undefined when $e = 0$.

The meaning of this definition may be made clearer by an example. Suppose v is in $\mathscr{C}[0, 100]$ and let

$$w = \int_0^{100} v(t)\, dt.$$

As $w \in \mathbb{R}$, the norm, $\|w\|$, is simply $|w|$. Now suppose the value $v(t)$ is subject to error $e(t)$. The maximum of $|e(t)|$ is $\|e\|$. Thus the entire graph of the error can be drawn on a strip of paper as shown in Figure 21. The error in w is $\int_0^{100} e(t)\, dt$. The worst possible case is when $e(t)$ has its maximum value throughout, or alternatively, when it is of maximum size but negative throughout. Either way, the error in w will have magnitude $100\|e\|$. So, if we are unlucky, the error may be multiplied by 100. If we write $w = Av$, then $\|A\| = 100$.

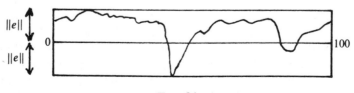

$\|e\|$

$\|e\|$

0

100

FIG. 21

The relevance of norms to the behaviour of errors has by now been adequately stressed. It will shorten explanations a little if, from now on, we simply discuss how vectors may be enlarged (or diminished) by linear operators, without insisting that such vectors be thought of as errors.

There is a way of formulating the definition of $\|A\|$ that is particularly convenient for graphical representation. This depends on the fact that the ratio $\|Av\|/\|v\|$ is unaltered if we replace v by kv, where k is any positive number. If v is any vector, other than 0, the number k can be chosen so that $\|kv\| = 1$, for $\|kv\| = k\|v\|$, so $k = 1/\|v\|$ does the trick.

Accordingly, there is no loss of generality if we consider the ratio $\|Au\|/\|u\|$ only for unit vectors, u. Then $\|u\| = 1$, so the ratio reduces simply to $\|Au\|$ and our definition becomes

$$\|A\| = \sup\{\|Au\| : \|u\| = 1\}.$$

This definition may be visualized as follows; take in turn all the points u of the unit sphere, $S(O, 1)$. Apply the operator A to each and consider the curve or surface formed by all the points Au. The radius, r, of the smallest sphere, $S(O, r)$, that contains all these points is the value of $\|A\|$.

Exercise. In the plane, $S(O, 1)$ is a circle when the ℓ_2 metric is used, a square with horizontal and vertical sides when the ℓ_∞ metric is used, and a tilted square when the ℓ_1 metric is used. Find and draw the figures to which these 'spheres' are mapped when the matrix

$$A = \begin{pmatrix} 1 & 1 \\ 0 & 2 \end{pmatrix}$$

acts on them. For each space, draw the appropriate $S(O, r)$ which is just large enough to contain all the output points. Deduce the value of $\|A\|$ for each of the three cases.

It should be noted that norm signs occur with three different meanings when operator norms are being discussed, at any rate in the general case. When we write $\|A\| = \sup\{\|Au\|_{\mathscr{Y}} : \|u\| = 1\}$ for a linear operator A, $\mathscr{X} \to \mathscr{Y}$, the symbol $\|u\|$ indicates the size of the vector u in the input space, $\mathscr{X}$, while $\|Au\|$ is the size of the output vector in the space, $\mathscr{Y}$. Finally $\|A\|$ involves a norm sign, which does not relate either to the space $\mathscr{X}$ or the space $\mathscr{Y}$, but is concerned with the effect of a mapping from one to the other.

Exercises on operator norms

1. Let $w = 2x - 3y + 4z$. Which point of the form $(\pm 1, \pm 1, \pm 1)$ makes w largest? If f is the function $\mathbb{R}^3 \to \mathbb{R}$, $(x, y, z) \to w$, what is $\|f\|_\infty$?

2. Generalize from your answer to question 1. What is $\|f\|_\infty$ for $f:(x, y, z) \to ax + by + cz$?

3. What is $\|f\|_\infty$ for

$$f:(x_1, x_2 \ldots x_n) \to \sum_1^n a_r x_r?$$

4. Let

$$w_1 = 2x_1 - 3x_2 + 4x_3,$$
$$w_2 = x_1 + x_2 + x_3.$$

Consider $f: x \to w$, where $x = (x_1, x_2, x_3)$ and $w = (w_1, w_2)$. What x with $\|x\|_\infty = 1$ makes $\|w\|_\infty$ a maximum? What is $\|f\|$, it being understood the ℓ_∞ norm applies both to input and output?

5. Would the answer to question 4 be different if, the rest of the equation being unchanged, w_2 were altered (a) to $x_1 + 2x_2 - 4x_3$, or (b) to $x_1 + 3x_2 - 6x_3$?

6. Investigate the generalization of the problems posed in questions 4 and 5, with the aim of finding a formula for $\|f\|$ where $f: x \to w$ is specified by

$$w_r = \sum_s a_{rs} x_s,$$

with $1 \leqslant r \leqslant m$ and $1 \leqslant s \leqslant n$.

7. Let $w = 2x - 3y + 4z$. What is the maximum value $|w|$ can have subject to $|x| + |y| + |z| = 1$? What is $\|f\|$ for $f:(x, y, z) \to w$, if the ℓ_1 norm is used for the input?

8. Generalize from your answer to question 7. What is $\|f\|$ for $f:(x, y, z) \to ax + by + cz$, the ℓ_1 norm applying to the input?

9. What is $\|f\|$ for

$$(x_1, x_2, \ldots, x_n) \to \sum_1^n a_r x_r,$$

the ℓ_1 norm applying to the input?

10. Let

$$w_1 = 7x_1 + 2x_2$$
$$w_2 = -3x_1 + 6x_2.$$

What is the maximum value of $|w_1| + |w_2|$ subject to $|x_1| + |x_2| = 1$? For $f : x \rightarrow w$, where $x = (x_1, x_2)$ and $w = (w_1, w_2)$ what x with $\|x\|_1 = 1$ makes $\|w\|_1$ a maximum? With these norms for x and w, what is $\|f\|$?

11. How would the answer to question 10 have to be modified if, the rest of the equation remaining unaltered, the equation for w_2 was changed to

(a) $w_2 = -3x_1 + 9x_2$?
(b) $w_2 = -3x_1 + 8x_2$?
(c) $w_2 = 3x_1 + 8x_2$?

It is instructive to show on graph paper the region to which f maps the unit ball in these various cases.

12. Investigate the general question of a formula for $\|f\|$ with f; $\ell_1 \rightarrow \ell_1$, $x \rightarrow w$, where

$$w_r = \sum_s a_{rs} x_s,$$

with $1 \le r \le m$ and $1 \le s \le n$.

13. If

$$c = \int_0^1 (3x^2 + 2x + 1) f(x) \, dx,$$

what is the maximum value c can take for a continuous function f subject to $|f(x)| \le 1$ for $0 \le x \le 1$?

14. What function, *not necessarily continuous*, makes c maximum, where

$$c = \int_0^{2\pi} f(x) \sin x \, dx \quad \text{and} \quad |f(x)| \le 1$$

for $x \in [0, 2\pi]$? If we also require f to be continuous what is the supremum for c? Is there any continuous f that makes c actually equal to the supremum value?

15. Discuss the general question of $\|T\|$, where T is in $\mathscr{C}[a, b]$ and T maps $f \rightarrow c$, where

$$c = \int_a^b \phi(x) f(x) \, dx$$

for a given continuous function ϕ. Question 13 throws some light on the situation when $\phi(x)$ is positive throughout, and question 14 on the case when $\phi(x)$ changes sign within the interval.

16. Let $T : \mathscr{C}[0, 1] \rightarrow \mathscr{C}[0, 1]$, $f \rightarrow g$ be defined by

$$g(x) = \int_0^1 (x^2 + 2xy + 3y^2) f(y) \, dy.$$

What f, with $\|f\| = 1$, makes $\|g\|$ maximum?, What is $\|T\|$?

17. (i) Let k be a prescribed number, for which $0 \le k \le 1$. Find the supremum of $|c|$, where

$$c = \int_0^1 (k - y)f(y)\, dy$$

and $f \in \mathcal{C}[0, 1]$ with $\|f\| = 1$. What values of k, in the interval $[0, 1]$ make sup $|c|$ a maximum, and what is the value of this maximum?

(ii) Let

$$g(x) = \int_0^1 (x - y)f(y)\, dy.$$

What is $\|T\|$ for $T: \mathcal{C}[0, 1] \to \mathcal{C}[0, 1], f \to g$?

18. Investigate $\|T\|$ for $T: \mathcal{C}[0, 1] \to \mathcal{C}[0, 1], f \to g$, where

$$g(x) = \int_0^1 K(x, y)f(y)\, dy,$$

the kernel $K(x, y)$ being continuous.

Note. If $K(x, y)$ were given by a table, for example by its values when x and y are both multiples of 0.1, an approximate treatment would lead us to consider equations of the type

$$g_r = \sum_s k_{rs} f_s.$$

Thus it is likely that there will be some resemblance between the answer to this question and the answer to question 6 above.

5.3. The space of bounded linear operators

Nothing said in the previous section excluded the possibility that an operator might produce an indefinitely large magnification. For an example of such an operator we need look no further than the familiar operation of differentiation, $D = d/dx$, applied to polynomials. Consider polynomial functions on the interval $[0, 1]$ with the $\mathcal{C}[0, 1]$ norm. Let $f(x) = x^n$. Then $Df(x) = f'(x) = nx^{n-1}$. Both of these take their maximum value for $x = 1$, so $\|f\| = 1$ and $\|f'\| = n$. The magnification, $\|Df\|/\|f\|$, is n, and as n can be arbitrarily large, there is no finite ceiling that can be put over the magnification.

A similar example can be constructed using trigonometric functions in $[0, \pi]$. If $f(x) = \sin nx$, $f'(x) = n \cos nx$, and once again $\|Df\|/\|f\| = n$.

An operator such as D is called *unbounded*. The operation of integration, as will be realized from the exercises at the end of section 5.2, is bounded; this makes integration much easier to study. This is the underlying reason why, in a book such as Hilbert and Courant, *Methods of Mathematical Physics*, the strategy throughout is to convert differential equation problems into equivalent problems with integral equations. In elementary calculus, differentiation is much easier to carry out than integration; this tends to obscure the fact that the theory of integral equations is much simpler than that of differential equations.

Accordingly, we turn our attention to bounded linear operators. The operator norm, defined in section 5.2, attaches a finite number, $\|A\|$, to each such operator A. This procedure does far more than allow us to estimate the magnification of errors. It turns out that bounded linear operators with this norm satisfy the vector axioms, V.1 to V.10, and the norm axioms, N.1 to N.5, and thus constitute a normed vector space. This means that we can apply to them the machinery already developed for such spaces; we can form infinite series of bounded linear operators, apply Cauchy tests to these, and establish convergence by considering the lengths of links in chains of operators.

The proof that all the norm and vector axioms are satisfied would make rather dull reading. It is much less boring, more instructive and a good test of whether the ideas have been digested, to work the proof out for yourself. In this work, very often a property of a bounded linear operator, $\mathscr{X} \to \mathscr{Y}$, is proved by appealing to the corresponding property of the spaces $\mathscr{X}$ and $\mathscr{Y}$. As a sample, suppose we wish to prove N.5, the norm version of the triangle inequality; it states $\|A + B\| \leq \|A\| + \|B\|$ for operators A and B, Let $\|A\| = a$ and $\|B\| = b$. We want to prove $\|A + B\| \leq a + b$. That is, if $u \in \mathscr{X}$ and $\|u\| = 1$, we want to prove $\|(A + B)u\| \leq a + b$. By the definition of $A + B$, this means $\|Au + Bu\| \leq a + b$. Now Au and Bu are vectors in the space $\mathscr{Y}$, which we know is a space meeting all the requirements of a normed, vector space. So, by N.5 for the space $\mathscr{Y}$, we have $\|Au + Bu\| \leq \|Au\| + \|Bu\|$. Now $\|Au\| \leq \|A\| \|u\| = a$, since $\|A\| = a$ and $\|u\| = 1$. Similarly, $\|Bu\| \leq b$. This gives us the result we want.

It may seem that this is cheating, for we seem to be proving axiom N.5 by appealing to N.5. But it is not so; that $\mathscr{X}$ and $\mathscr{Y}$ are normed, vector spaces is part of the data. If this were not given, the assertion about the operators $\mathscr{X} \to \mathscr{Y}$ could well be false or meaningless. Our problem is to show from the known properties of the spaces $\mathscr{X}$ and

$\mathcal{Y}$, that similar properties hold for a new object, the space consisting of bounded, linear operators, $\mathcal{X} \to \mathcal{Y}$.

The symbol $\mathcal{B}(\mathcal{X}, \mathcal{Y})$ is used for the space of bounded linear operators, $\mathcal{X} \to \mathcal{Y}$. There is no danger of confusion with the symbol for the open ball, $B(v, r)$, in which a vector and a number appear, not a pair of spaces.

One further property of $\mathcal{B}(\mathcal{X}, \mathcal{Y})$ should be mentioned. It can be shown that, if $\mathcal{Y}$ is complete, then $\mathcal{B}(\mathcal{X}, \mathcal{Y})$ also is complete. This means that if we have a sequence of bounded, linear operators, satisfying the Cauchy condition, then the limit of this sequence will exist, and will belong to the space $\mathcal{B}(\mathcal{X}, \mathcal{Y})$; that is to say, it also will be a bounded, linear operator. In all the examples we consider, $\mathcal{X}$ and $\mathcal{Y}$ will both be Banach spaces, so it is sufficient to remember that in this case $\mathcal{B}(\mathcal{X}, \mathcal{Y})$ will also be a Banach space.

In matrix theory, square matrices, M, are particularly interesting, since for them we can form the powers $M^2, M^3, \ldots$ This possibility exists because a square matrix arises when we map a space to itself, so the output vector, Mv, is in the input space and we can apply the operator M to it again and get $MMv = M^2v$. The same possibility exists for bounded, linear operators $\mathcal{X} \to \mathcal{X}$. The space of such operators is written $\mathcal{B}(\mathcal{X}, \mathcal{X})$, sometimes abbreviated to $\mathcal{B}(\mathcal{X})$. Sometimes $\mathcal{L}(\mathcal{X}, \mathcal{X})$ is used instead of $\mathcal{B}(\mathcal{X}, \mathcal{X})$, the $\mathcal{L}$ standing for 'linear'.

In this connection it should be noted that some books, in particular those translated from the Russian, apply the term *linear operator* only to those operators we have called *linear, bounded*. What we call 'linear', they call 'additive and homogeneous'. An operator A is additive if, for all vectors u, v, the equation $A(u + v) = Au + Av$ is valid. It is homogeneous if $A(ku) = k(Au)$ for any vector u and any number k. The terms 'additive' and 'homogeneous' are in general use. The Russians differ only in requiring boundedness, as well as the other two properties, before they permit an operator to be called linear.

In reading the literature you may well be puzzled as to the shades of meaning that differentiate such words as function, mapping, correspondence, transformation and operator. The first four need cause no trouble. In the past they may have had different associations; today they seem to be used interchangeably. Operator still carries a separate meaning. Operators of course are functions. The distinction comes when we introduce norms. If we define $\|M\|$ as the supremum of $\|Mu\|$ for $\|u\| = 1$, we are using the operator norm. By speaking of M as an operator, we are giving notice that we intend to use this norm.

In section 5.4, we shall consider operator norms related to ℓ_∞, ℓ_1 and $\mathscr{C}[a, b]$. These are algebraically simple to handle. We defer consideration of ℓ_2 and general ℓ_p norms, since these require certain inequalities, which will have a chapter to themselves, namely chapter 9.

Exercises

1. Let T; $\mathscr{C}[0, 1] \to \mathscr{C}[0, 1]$ be defined by $Tf = g$ where

$$g(x) = \int_0^x tf(t)\, dt.$$

Find T and discuss the convergence of the series $I + T + T^2 + \ldots + T^n + \ldots$ in the space of bounded operators.

Examine the series produced by the iteration

$$f_{n+1}(x) = 1 + \int_0^x tf_n(t)\, dt \quad \text{with} \quad f_0(x) \equiv 0.$$

Identify the analytic function given by this series. Is it a solution of the integral equation

$$f(x) = 1 + \int_0^x tf(t)\, dt?$$

2. Let matrix

$$M = \begin{pmatrix} 0 & a \\ b & 0 \end{pmatrix}.$$

Find the matrix N given by the infinite series $N = I + M + M^2 + \ldots + M^n + \ldots$, when this series converges. What condition must a and b satisfy if the series that appear in the four entries of N are all to converge? Find algebraic expressions for the entries of N when this condition is satisfied.

On the analogy of results in elementary algebra, it appears plausible that $(I - M)N = I$. Is this equation in fact verified?

What are the eigenvalues of M?

3. Let M; $\mathscr{C}[0, 1] \to \mathscr{C}[0, 1]$ be defined by $Mf = g$, where

$$g(x) = \int_0^x yf(y)\, dy + \int_x^1 xf(y)\, dy.$$

Find $\|M\|$. Will the iteration defined by $f_{n+1} = 1 + Mf_n$ converge? Here **1** denotes the function with the constant value 1, and we suppose $f_0(x) \equiv 0$.

5.4. Some operator norms

The exercises at the end of section 5.2 provided exploratory material for the finding of various operator norms. We now summarize the results to which these explorations could lead.

1. $\ell_\infty \to \mathbb{R}$. We require the maximum of $|w|$, where $w = \sum_1^n a_r x_r$ and $\|x\|_\infty = 1$. We have

$$|w| = \left| \sum_1^n a_r x_r \right| \leq \sum_1^n |a_r| \, |x_r| \leq \sum_1^n |a_r|$$

since $|x_r| \leq 1$ for each r. We can make $w = \sum_1^n |a_r|$ by taking $x_r = $ sign a_r, that is to say, by taking $x_r = +1$ when a_r is positive and $x_r = -1$ when a_r is negative. It does not matter what x_r we choose if $a_r = 0$. Accordingly for the function $\phi : x \to w$ we have $\|\phi\|_\infty = \sum_1^n |a_r|$.

2. $\ell_\infty \to \ell_\infty$. We now consider ϕ; $x \to w$ with

$$w_r = \sum_{s=1}^n a_{rs} x_s, \quad \text{for} \quad r = 1, \ldots, m.$$

Now $\|w\| = \max |w_r|$. So we consider each $|w_r|$ in turn, see how large we can make it, and choose the largest of these. In problem (1) above, we saw that we could choose x, with $\|x\| = 1$, to make $|w_r| = \sum_{s=1}^n |a_{rs}|$. Choosing r so as to get the largest of these, we see that $\|w\| = \max_r \sum_{s=1}^n |a_{rs}|$, so this is the value of $\|\phi\|_\infty$.

For example, in question 5(b) of the exploratory exercises, we had $w_1 = 2x_1 - 3x_2 + 4x_3$ and $w_2 = x_1 + 3x_2 - 6x_3$. The maximum value of w_1 is 9, obtained from $x = (1, -1, 1)$, while the maximum value of w_2 is 10, from $x = (1, 1, -1)$. The larger number is 10 so $\|\phi\|_\infty = 10$.

3. $\ell_1 \to \mathbb{R}$. As in (1), $w = \sum_1^n a_r x_r$, but now $\sum_1^n |x_r| = 1$ is the condition. It now pays to concentrate our resources on the largest $|a_r|$. For instance, if $w = 3x_1 + 10x_2 + 2x_3$, any investment in x_2 is multiplied by 10: it would be unwise to put any of our unit total into x_1 or x_3.

Thus $x = (0, 1, 0)$ gives the maximum value, 10, for w. As absolute values are involved, signs can be neglected. Thus, if $w = 3x_1 - 10x_2 + 2x_3$, then $w = 10$ is achieved by $x = (0, -1, 0)$ with $\sum_1^3 |x_r| = 1$. Quite generally, the maximum $|w|$ is max $|a_r|$, and this is $\|\phi\|_1$. This can be proved formally.

4. $\ell_1 \to \ell_1$. We now wish to maximize

$$\sum_1^m |w_r| \quad \text{for} \quad w_r = \sum_{s=1}^n a_{rs} x_s \quad \text{with} \quad \sum_1^n |x_s| = 1,$$

as in questions (10) to (12) of the exploratory exercises.

This question is probably less easy than the others we have considered, so it is worth while to look at the simple example of question 10 which has

$$w_1 = \quad 7x_1 + 2x_2$$
$$w_2 = -3x_1 + 6x_2.$$

As in problem 1, we can obtain inequalities which show that $|w_1| + |w_2|$ cannot exceed a certain number, which we then show can be attained. We have

$$|w_1| + |w_2| = |7x_1 + 2x_2| + |-3x_1 + 6x_2|$$
$$\leqslant |7x_1| + |2x_2| + |-3x_1| + |6x_2|$$
$$= 7|x_1| + 2|x_2| + 3|x_1| + 6|x_2|$$
$$= 10|x_1| + 8|x_2|.$$

This last expression is like the expression we had just now in problem 3. As $|x_1| + |x_2| = 1$, it cannot exceed 10, and it equals 10 if $x_1 = 1$ and $x_2 = 0$. Going back to the original equations, and taking $x_1 = 1$, $x_2 = 0$, we find that $w_1 = 7$, $w_2 = -3$, and indeed $|w_1| + |w_2| = 10$.

Our questions may be put in matrix form as $w = Ax$, with

$$A = \begin{pmatrix} 7 & 2 \\ -3 & 6 \end{pmatrix}.$$

This helps to show the significance of the numbers 10 and 8 that occurred in the work above; 10 is the sum of the absolute values of the numbers in the first column of A, and 8 is the corresponding sum for the second column.

We can deal with the general problem, as posed at the beginning of this discussion, by exactly the same strategy. The conclusion is that for $\phi : x \rightarrow w$, we have

$$\|\phi\|_1 = \max_s \sum_{r=1}^m |a_{rs}|.$$

This is simply the algebraic form of the instruction that we are to find the sum of the absolute values of the elements in each column, and choose the largest such sum.

The thoughts leading to this algebraic argument are perhaps not too transparent. There is an interesting alternative treatment along geometrical lines. It uses convexity properties, and has some kinship to the argument in linear programming that maximum values occur at corners of the permitted region. The corners of $S(O, 1)$ in ℓ_1 of course are the points where all the co-ordinates except one are nought.

5. $\mathscr{C}[a, b] \rightarrow \mathbb{R}$. We now consider the problem stated in exploratory question 15; if

$$c = \int_a^b \phi(x)f(x)\, dx,$$

where ϕ is a given function continuous on $[a, b]$, what is the norm of $T : f \rightarrow c$?

Consider first the case where $\phi(x)$ is positive throughout. Since $\|f\| = 1$, $-1 \leq f(x) \leq 1$. It is pretty clear that if we want to make the integral as big as possible we should take $f(x) \equiv 1$.

Now suppose we have a case, like $\phi(x) = 1 - 2x$, where $\phi(x)$ is first positive and then negative. We would like to take $f(x) = 1$ for $x < 0.5$ and $f(x) = -1$ for $x > 0.5$. The only trouble is that f is then not continuous. Suppose we take $f(x) = 1$ for, say, $x \leq 0.499$ and $f(x) = -1$ for $x \geq 0.501$. In the interval $(0.499, 0.501)$ we let $f(x)$ rush down from $+1$ to -1. In this interval we could, for instance, define $f(x) = 1000(0.5 - x)$. The resulting value of c would be very nearly $\int_a^b |\phi(x)|\, dx$. We can get closer to this value by taking a quicker descent in a narrower interval. We can in fact get as close as we like to it, but never attain it. It is in view of such situations that norms are defined in terms of a supremum (a ceiling that may or may not be reached) rather than in terms of a maximum (a ceiling that is reached).

The device just explained can easily be extended to deal with the case where ϕ has a finite number of changes of sign in $[a, b]$. We can even save our conclusion in cases like $\phi(x) = x \sin(1/x)$, with an infinity of sign changes. Always

$$\|T\| = \int_a^b |\phi(x)| \, dx.$$

6. $\mathscr{C}[a, b] \to \mathscr{C}[a, b]$. We now consider $\|T\|$ for $T: f \to g$, with

$$g(x) = \int_a^b K(x, y)f(y) \, dy.$$

First of all note that this problem has an analogy with problem 2 above; x plays the role of r, while y plays the role of s, and integration replaces summation.

By definition, $\|T\|$ is the supremum of $\|g\|$, subject to the condition $\|f\| = 1$. In its turn, $\|g\|$ is the supremum of $|g(x)|$. We have two things at our disposal, the number x and the function f, and we want to choose these (subject to the limitations put on them) to make $|g(x)|$ as large as possible. We can suppose this done in two stages. First suppose x fixed. Then $K(x, y)$ depends only on y and we may call it $\phi(y)$, so that

$$g(x) = \int_a^b \phi(y)f(y) \, dy.$$

From problem 5 we know how to make the best choice of f for this, and we know it leads to the value $\int_a^b |\phi(y)| \, dy$, which is the same thing as $\int_a^b |K(x, y)| \, dy$. This would be the best we could do, if someone had fixed the value of x for us. But nobody has done this. Accordingly we are free to examine how the value of this integral depends on the choice of x, and choose that x which makes it as large as possible. (Note that the value of this integral depends only on the value of x. No other variable is involved.) Accordingly

$$\|T\| = \sup_{x \in [a, b]} \int_a^b |K(x, y)| \, dy.$$

5.5. Properties of operators

1. With operator norm defined, we can measure the distance between two operators, A and B, by the number $\|B - A\|$. This in turn

allows us to define the limit of a sequence of operators; we say $T_n \to T$ when $\|T_n - T\| \to 0$.

The business of operators is to act on vectors, so the question naturally arises; if $T_n \to T$, does it follow that $T_n v \to Tv$ for every vector v? It does, and the proof is simple. Suppose T_n and T are operators $\mathscr{X} \to \mathscr{Y}$. Three different norms are involved. To make this explicit, $\|\ \|_{\mathscr{X}}$ will be used for the norm of vectors in the input space $\mathscr{X}$, similarly $\|\ \|_{\mathscr{Y}}$ for vectors in the output space $\mathscr{Y}$, and $\|\ \|_{\mathcal{O}}$ for the norm of the operators. To see whether $T_n v$ tends to Tv, we have to consider $\|T_n v - Tv\|_{\mathscr{Y}}$. Now

$$\|T_n v - Tv\|_{\mathscr{Y}} = \|(T_n - T)v\|_{\mathscr{Y}} \leqslant \|T_n - T\|_{\mathcal{O}} \|v\|_{\mathscr{X}},$$

and this does tend to nought, since $\|v\|_{\mathscr{X}}$ is constant and the factor $\|T_n - T\|_{\mathcal{O}} \to 0$.

The converse of this theorem is not true. Even if $T_n v \to Tv$ for every vector v, it need not be the case that $\|T_n - T\| \to 0$.

2. If A and B are two operators for which the product AB is defined, the operator norm has a property that is very useful in calculations, namely $\|AB\| \leqslant \|A\| \cdot \|B\|$. This is almost obvious. When B acts, it multiplies the length, or norm, of a vector by $\|B\|$ at most. When A acts on the output Bv, it multiplies the length by $\|A\|$ at most. The combined effect of these operations cannot multiply length by more than $\|A\| \cdot \|B\|$.

If A happens to be an operator $\mathscr{X} \to \mathscr{X}$, iteration is possible and the powers $A^2, A^3 \ldots A^n \ldots$ are defined. The same argument applied to these shows $\|A^2\| \leqslant \|A\|^2$ and generally $\|A^n\| \leqslant \|A\|^n$.

As a rule, $\|A^2\|$ is actually less than $\|A\|^2$. A simple example is given by the matrix

$$A = \begin{pmatrix} 0 & 3 \\ 2 & 0 \end{pmatrix}$$

in the traditional setting, that is to say, in Euclidean geometry as described with squared graph paper. Then

$$A^2 = \begin{pmatrix} 6 & 0 \\ 0 & 6 \end{pmatrix},$$

from which it follows that $\|A\| = 3$ and $\|A^2\| = 6$, which is less than 3^2. The reason is that A trebles the length of the vector $(0, 1)$ when it acts, but it sends this vector to (3, 0), which only gets doubled in length when A acts again. There is no vector whose length gets multiplied by 3 at both stages of the operation A^2. That equality can happen is shown by the example,

$$B = \begin{pmatrix} 3 & 0 \\ 0 & 2 \end{pmatrix},$$

for which $\|B\| = 3$ and $\|B^2\| = 9$.

3. We shall have to deal with many operators defined by limits. If T is defined as $\lim T_n$, we may meet expressions involving T, such as $A + T$ or AT. Are we entitled to assume $A + T = \lim(A + T_n)$ and $AT = \lim(AT_n)$?

The first question amounts to asking, does $A + T_n$ approach $A + T$? The distance between these two operators is $\|(A + T_n) - (A + T)\|$, which simplifies to $\|T_n - T\|$ and approaches nought since $T_n \to T$. The answer to the first question is 'yes'.

The second question involves the distance of AT_n from AT. As $AT_n - AT = A(T_n - T)$, we are dealing with the product of two operators. We can make use of the result proved in item 2 just now, that the norm of a product does not exceed the product of the norms. So $\|A(T_n - T)\| \leqslant \|A\| \|T_n - T\|$. Here we are assuming that A is a bounded, linear operator. The final expression clearly does approach nought, so it is true that $AT_n \to AT$. A similar argument shows $T_n A \to TA$.

Naturally, there are other theorems about limits of operators that are needed to justify steps in computation. As a rule, they closely resemble theorems in classical analysis.

5.6. Applications of operator norms

We now look at some examples of the use that can be made of operator norms in numerical analysis.

Many problems take the form of finding the vector v for which $Av = b$, where A is a linear operator and b is a known vector. A familiar, simple case is when A is a matrix with n rows and n columns. If A has an inverse, the solution is $v = A^{-1}b$. So naturally the questions are—does A have an inverse?—if there is an error in b, how does this affect the solution v?—if A itself is only approximately correct, what effect does this have on v?—what happens if both sources of error are present?

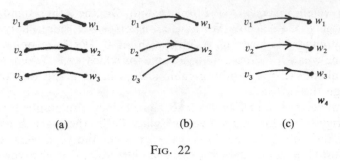

(a)   (b)   (c)

FIG. 22

Figure 22 shows diagrammatically the situations in which there is, or is not, an inverse. In situation (a) there is a one-to-one correspondence between inputs and outputs. The inverse function, $A^{-1}: w \to v$, is defined, since to each output, w, of A there corresponds just one input, v. In situation (b), $Av_2 = Av_3 = w_2$ and we cannot define A^{-1}, since we do not know whether A^{-1} should map w_2 to v_2 or to v_3. If A is linear, then, in situation (b), $A(v_2 - v_3) = Av_2 - Av_3 = 0$, but, as $v_2 \neq v_3$, the input $v_2 - v_3 \neq 0$. Thus there is a useful test for linear operators; no inverse of A exists if, and only if, $Av = 0$ has a solution v other than $v = 0$.

In situation (c) the position is rather different. The correspondence is one-to-one, except for the existence of a potential output, w_4, that does not arise for any input. In this situation, we can to some extent rescue the inverse by defining an inverse A^{-1} for the outputs w_1, w_2, w_3 only.

As an example of situation (c), consider the operations of differentiation and integration, which obviously have something resembling an inverse relationship. Let D denote the operator d/dx and T the operator $f \to g$ with

$$g(x) = \int_0^x f(t)\, dt.$$

Whether D and T are inverses depends on the vector spaces for which we suppose them defined.

In order to avoid all analytic complexities, we will suppose that T is regarded as an operator $\mathcal{X} \to \mathcal{Y}$, where $\mathcal{X}$ consists of all quadratics $ax^2 + bx + c$, and $\mathcal{Y}$ of all cubics, $px^3 + qx^2 + rx + s$. Then D is an operator $\mathcal{Y} \to \mathcal{X}$, but D^{-1} does not exist, since we cannot determine the constant term, s, in the cubic by examining the quadratic that arises when the cubic is differentiated. On the other hand, the operator $T: \mathcal{X} \rightarrow \mathcal{Y}$ does not have an inverse, T^{-1}, since applying T to any quadratic leads to a cubic with $s = 0$. We have no way of defining what T^{-1} should do to a cubic for which $s \neq 0$. When T acts on a quadratic, the resulting cubic always has a graph that passes through the origin.

However, we can effect a rescue, since cubics with graphs through the origin do in fact form a vector space. So let the space $\mathcal{X}$ consist of all the cubics of the form $px^3 + qx^2 + rx$, and regard D as an operator $\mathcal{X} \rightarrow \mathcal{X}$, and T as $\mathcal{X} \rightarrow \mathcal{X}$. Then D and T are each inverses of the other. For $D(px^3 + qx^2 + rx) = 3px^2 + 2qx + r$, which will be the same as the quadratic $ax^2 + bx + c$ if $a = 3p$, $b = 2q$, $c = r$. These equations clearly define a one-to-one relation between inputs and outputs.

In reading the literature, some care is needed in interpreting an author's statement, 'A has no inverse'. In some contexts it may mean that either situation (b) or situation (c) exists; in other contexts it may mean that we definitely have situation (b). Some confusion can arise if this is not borne in mind.

Much work in analysis is related to the traditional formula

$$\frac{1}{1-x} = 1 + x + x^2 + x^3 + \ldots + x^n + \ldots .$$

This may be derived in various ways—by long division, 1 being divided by $1-x$; by summing the geometrical progression on the right-hand side; by applying the binomial theorem to $(1-x)^{-1}$ and simplifying; by iterating to solve $y = 1 + xy$. This is a simple formula, but it should be in a state of instant recall and instant recognition, for it is of use in many problems.

One of its traditional uses was for finding the numerical values of reciprocals. For example, by putting $x = -0.005$ we have

$$1/1.005 = 1 - 0.005 + 0.000\,025 - \ldots$$

from which a result to many places of decimals can be found very quickly. Such use was not restricted to numbers near to 1. To find $1/4.02$ we observe $4.02 = 4(1.005)$, so a quarter of the series above gives this reciprocal.

This approach can be generalized to much more complicated problems, such as finding the inverse of a matrix or some other bounded, linear operator.

The series for $1/(1-x)$ is convergent for $|x| < 1$. It is not surprising to find that the series for $(I-T)^{-1}$ requires the condition $\|T\| < 1$. Here I stands for the identity operator, $v \to v$.

THEOREM 1.   If $T \in \mathcal{B}(\mathcal{X}, \mathcal{X})$ and $\|T\| < 1$, then $(I-T)^{-1}$ exists and is given by the convergent series $I + T + T^2 + \ldots + T^n + \ldots$.

The proof consists in comparing the norms of the terms in this series with the terms of the convergent geometrical progression $1 + k + k^2 + \ldots$, where $k = \|T\|$. This shows the series to be convergent. We must also show that it converges to the correct reciprocal. Let S represent the sum of the series. We want to show $S(I-T) = (I-T)S = I$. Consider the sum of the first n terms, $S_n = I + T + T^2 + \ldots + T^{n-1}$. By the usual processes of algebra we see that $S_n(I-T) = (I-T)S_n = I - T^n$. As $\|T^n\| \le k^n$ and $k^n \to 0$, it follows that $T^n \to 0$. Letting n tend to infinity, we obtain the equations we want. If this argument is studied in detail, it will be seen that we are using the properties of limits established in item 3 at the end of section 5.5.

We assume, as we shall do throughout, that $\mathcal{X}$ is a Banach space.

From the same series we can obtain an estimate for the norm of the inverse.

For

$$\|S_n\| = \|I + T + T^2 + \ldots T^{n-1}\|$$
$$\le \|I\| + \|T\| + \|T^2\| + \ldots + \|T^{n-1}\|$$
$$\le 1 + k + k^2 + \ldots + k^{n-1} < 1/(1-k).$$

As $S_n \to S$, it seems reasonable, on geometrical grounds, to expect $\|S_n\| \to \|S\|$. This will be proved in section 9.3 with the label 'continuity of the norm'. For the present we assume it, and also that the limit operator S will be linear, and state Theorem 2.

THEOREM 2.   If $\|T\| = k < 1$, the inverse $S = (I-T)^{-1}$ exists and is a bounded, linear operator with $\|S\| \le (1-k)^{-1}$.

We now take a very natural step, corresponding to the arithmetical illustration at the outset where finding $1/4.02$ was reduced to finding $1/1.005$. If A is a bounded, linear operator with the inverse A^{-1}, then $B = A(I-T)$ has the inverse $B^{-1} = (I-T)^{-1}A^{-1}$. Now $B = A - AT = A - C$, where $C = AT$ and $T = A^{-1}C$. If B is some operator fairly close to A, the difference $C = A - B$ is the most natural thing to consider. We would rather have a condition on C than the condition on T that figured in Theorems 1 and 2. As $\|T\| = \|A^{-1}C\| \leqslant \|A^{-1}\| \|C\|$, we shall be sure that $\|T\|$ is less than 1 if we stipulate $\|A^{-1}\| \|C\| < 1$. So we have Theorem 3.

THEOREM 3. If a bounded, linear operator A has an inverse, A^{-1}, and $B = A - C$, where $\|C\| < 1/\|A^{-1}\|$, then B^{-1} exists and is a bounded, linear operator.

The next thing we want to know is how much B^{-1} is going to differ from A^{-1} as the result of this change C away from A. The series we have been considering leads us quite naturally to an answer.

$$
\begin{aligned}
B^{-1} - A^{-1} &= (I - T)^{-1}A^{-1} - A^{-1} \\
&= (I + T + T^2 + \ldots)A^{-1} - A^{-1} \\
&= (T + T^2 + \ldots)A^{-1} = (I + T + T^2 + \ldots)TA^{-1} \\
&= (I - T)^{-1}TA^{-1}.
\end{aligned}
$$

Actually, this equation can be derived by algebra from the equation $B^{-1} = (I - T)^{-1}A^{-1}$, but it is then probably less clear how the idea was arrived at.

If we now take norms of both sides and use the inequality stated in Theorem 2, that $\|(I - T)^{-1}\| \leqslant 1/(1 - \|T\|)$, we reach Theorem 4.

THEOREM 4. If $B = A - C$, and $\|C\| < 1/\|A^{-1}\|$, then

$$
\|B^{-1} - A^{-1}\| \leqslant \frac{\|T\|}{1 - \|T\|} \cdot \|A^{-1}\|
$$

where $T = A^{-1}C$.

If we replace $\|T\|$ by an overestimate of $\|T\|$, the fraction $\|T\|/(1 - \|T\|)$ will be replaced by a larger number. As we know that $\|A^{-1}\| \|C\|$ is such an overestimate, we can get a result in which T does not appear, but only A, B and C, namely Theorem 5.

THEOREM 5. If $B = A - C$ and $\|C\| < 1/\|A^{-1}\|$, then B^{-1} exists and

$$
\|B^{-1} - A^{-1}\| \leqslant \frac{\|A^{-1}\|^2 \|C\|}{1 - \|A^{-1}\| \|C\|}.
$$

These theorems are useful in numerical analysis. For instance, if we have a system of equations, $Av = u$, and we solve them by

Gaussian elimination, the effect of rounding-off is to give the solution of a different system, $Bv = u$. A rule for estimating $\|C\|$ is known. (See I. Gladwell in chapter 3 of Delves and Walsh.) Now let $v_0 = A^{-1}u$ represent the true solution and $v_1 = B^{-1}u$ the solution obtained from the equations affected by round-off error. Then

$$\|v_1 - v_0\| = \|B^{-1}u - A^{-1}u\| = \|(B^{-1} - A^{-1})u\| \leqslant \|B^{-1} - A^{-1}\| \, \|u\|.$$

Accordingly, by using Theorem 5, we can immediately estimate the size of the error.

Often we are interested not in the actual size of the error, but rather in the ratio of the error to the quantity studied, here given by the number $\|v_1 - v_0\|/\|v_0\|$. By arguments resembling those already used, it can be shown that this ratio does not exceed $\|A^{-1}\|\,\|C\|/(1 - \|A^{-1}\|\,\|C\|)$. This last result in fact is the one I. Gladwell was concerned with, in the chapter just cited.

It is to be noted that, in these formulas, $1/\|A^{-1}\|$ cannot be replaced by any simpler expression. It is *not*, for instance, the same as $\|A\|$, as one might guess on the analogy with real numbers. For instance, if we take for A the matrix

$$\begin{pmatrix} 2 & 0 \\ 0 & 0.1 \end{pmatrix}$$

then

$$A^{-1} = \begin{pmatrix} 0.5 & 0 \\ 0 & 10 \end{pmatrix}.$$

Then $\|A\| = 2$ and $\|A^{-1}\| = 10$. It is clear no functional relationship can connect these two.

The nature of the inverse matrix, A^{-1}, is a matter that has to be considered very carefully whenever we are solving the system of equations, $Av = u$, for $v = A^{-1}u$, and if the matrix for A^{-1} involves large numbers, small errors in u may produce large errors in v. It is an unfortunate fact that this ill-conditioning occurs in the oldest method of approximating a function by a polynomial, the method of least squares, which will be discussed in chapter 8. If, for example, we seek the fourth-degree polynomial that fits a given function as closely as possible on the interval $[0, 1]$, the method of least squares leads to equations for the coefficients involving the matrix, A, which is

$$
\begin{pmatrix}
1 & \frac{1}{2} & \frac{1}{3} & \frac{1}{4} & \frac{1}{5} \\
\frac{1}{2} & \frac{1}{3} & \frac{1}{4} & \frac{1}{5} & \frac{1}{6} \\
\frac{1}{3} & \frac{1}{4} & \frac{1}{5} & \frac{1}{6} & \frac{1}{7} \\
\frac{1}{4} & \frac{1}{5} & \frac{1}{6} & \frac{1}{7} & \frac{1}{8} \\
\frac{1}{5} & \frac{1}{6} & \frac{1}{7} & \frac{1}{8} & \frac{1}{9}
\end{pmatrix}
$$

The inverse, A^{-1}, is the matrix

$$
\begin{pmatrix}
25 & -300 & 1050 & -1400 & 630 \\
-300 & 4800 & -18900 & 26880 & -12600 \\
1050 & -18900 & 79380 & -117600 & 56700 \\
-1400 & 26880 & -117600 & 179200 & -88200 \\
630 & -12600 & 56700 & -88200 & 44100
\end{pmatrix}
$$

In the worst possible case, if the vector u happened to be in error by $(k, -k, k, -k, k)$, the resulting error in the fourth coefficient would be $413280k$. When polynomials of higher degree are fitted, even larger numbers occur. The matrices are known as Hilbert matrices, and data on them will be found in J. Todd, (1).

As a rule, it is laborious to calculate A^{-1} for a given matrix or linear operator. The moral to be drawn from the example above is not that we should always calculate $\|A^{-1}\|$, which would be very time-consuming, but rather that we should always be aware of the possibility that $\|A^{-1}\|$ may be large, and take whatever precautions seem appropriate. Sometimes, of course, we can estimate $\|A^{-1}\|$ without difficulty, as for instance when $A = I - T$ by using Theorem 2 of this section.

A useful source of further information is Delves and Walsh. In chapter 13, G. F. Miller's discussion of integral equations is dominated by the question of ill-conditioning. In example 3 of chapter 8, section 5, I. Barrowdale gives a very striking example of ill-conditioning, in which wildly differing solutions are obtained to the same problem, depending only on whether the initial data are given to three, six or twelve places of decimals. In chapter 9, section 5(c), L. M. Delves gives a formula for estimating the error in v, the solution of $Av = u$, when both A and u are subject to error.

5.7. Continuity of bounded operators

If we have the iteration defined by $v_{n+1} = f(v_n)$, and find $v_n \to v$, we know that $v_{n+1} \to v$, and at first sight it may seem justified to jump to the conclusion $v = f(v)$. However an example will show that this involves a certain assumption about the nature of f.

Consider $f : \mathbb{R} \to \mathbb{R}$, defined by $f(x) = 0.5x$ for $x \neq 0$ and $f(0) = 1$. Now in fact $f(x) = x$ has no solution, for if $x \neq 0$, then $x \neq 0.5x$, and for $x = 0$, $f(0) = 1 \neq 0$. Whatever initial value x_0 is chosen, the iteration with $x_{n+1} = f(x_n)$ will involve repeated halving, if $x_0 \neq 0$, and in any case will make $x_n \to 0$. If in the equation $x_{n+1} = f(x_n)$ we let $n \to \infty$, we find $\lim x_{n+1} = \lim f(x_n)$, so $\lim f(x_n) = 0$, which indeed is so. To prove that $x = 0$ was a solution of $x = f(x)$, we would have to take the further step, $0 = \lim f(x_n) = f(\lim x_n) = f(0)$, but this would be justified only if f were continuous, which it is not.

Accordingly, if the iteration $v_{n+1} = f(v_n)$ makes $v_n \to v$, we can be sure that $v = f(v)$ only if we know that f is continuous. Fortunately, if f is a bounded, linear operator, this logical loophole does not cause any trouble, for we have Theorem 6.

THEOREM 6.   Every bounded, linear operator is continuous.

The proof is straightforward. Let A be a bounded, linear operator with $\|A\| = c$. If $Av_1 = w_1$ and $Av_2 = w_2$, then $\|w_1 - w_2\| = \|Av_1 - Av_2\| = \|A(v_1 - v_2)\| \leqslant \|A\| \|v_1 - v_2\| = c\|v_1 - v_2\|$. It is evident that we can make the change in the output as small as we like by taking a sufficiently small change in the input.

As usual, we may have a situation where the theorem in pure mathematics may not help in practical computation. For example, if $\|A\| = 10^{20}$, then $v \to Av$ is theoretically continuous, but for practical purposes it might almost as well be discontinuous.

The converse of Theorem 6 is also true. If a linear operator is continuous, it must be bounded.

6 Differentiation and integration

6.1. A classical theory of iteration

IN THIS CHAPTER we examine certain classical applications of calculus and consider whether these ideas can be generalized and given wider scope.

The first application we consider is the use of the derivative, f', to give information about the iteration with $x_{n+1} = f(x_n)$. If a is a fixed point of f, that is, if $f(a) = a$, and we take $x = a + h$, where h is small, then $f(x) = f(a+h) \simeq f(a) + hf'(a) = a + hf'(a)$. It seems plausible that, if $|f'(a)| < 1$, then $f(x)$ will be nearer to a than x was, while, if $|f'(a)| > 1$, $f(x)$ should be further away. Thus the size of $|f'(a)|$ is used to classify fixed points into attractive and repulsive types.

It is also true that, if all the points of a sequence $\{x_n\}$ lie in a region for which $|f'(x)| \leq k$ with $k < 1$, then $x \to f(x)$ will be a contraction operator in this region and the sequence will certainly converge. In fact, if b and c are two points in this region, we can prove $|f(c) - f(b)| \leq k |c - b|$. This is done with the help of the mean value theorem, which asserts that for some X between b and c we have $f(c) - f(b) = f'(X)(c - b)$. (This is equivalent, in Figure 23, to the assertion that, at some point Z in the arc BC, the tangent is parallel to the chord BC.) As we know $|f'(X)| \leq k$, the desired result follows immediately.

In applications we usually know some region for which $|f'(x)| \leq k$, and we need to check that the sequence remains in this region as the iteration proceeds. Using the image we had earlier, of a chain joining the points $x_0, x_1, x_2 \ldots$, we know that each link is at most k times as long as the previous one. Thus the total length of the chain cannot exceed $|x_1 - x_0| (1 + k + k^2 + \ldots) = |x_1 - x_0|/(1 - k)$. All will be well, provided all points within this distance of x_0 lie in the region.

For example, suppose we wish to solve $x = e^x - 1.1$. For $f(x) = e^x - 1.1$, $f'(x) = e^x$ and $|f'(x)| < 0.8$ if $x < -0.2232$. If we try $x_0 = -0.6$, we find $x_1 = -0.551\ 188$ so $|x_1 - x_0|/(1 - 0.8) < 0.05/0.2 = 0.25$, which is much less than the distance from -0.6 to -0.2232, the edge of the region, so our guess, that $|f'(x)| < 0.8$ remains valid throughout the iteration, is justified.

We can estimate the number of iterations required. The length of

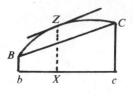

FIG. 23

the chain that comes after x_n does not exceed $\|x_1 - x_0\|(k^n + k^{n+1} + \ldots) = k^n \|x_1 - x_0\|/(1 - k)$, hence does not exceed $(0.8)^n(0.25)$. This gives an upper estimate for the distance of x_n from the limit point. We can make it less than 10^{-9} by taking $n = 87$. This estimate of course errs on the side of pessimism, since the graph of $e^x - 1.1$ gets steeper as x increases. Actually the iteration takes place between $x_0 = -0.6$ and $x = -0.483\,183$, with the slope nowhere more than about 0.62. The estimate of 87 iterations is the 'a priori' estimate—the estimate that can be made before any calculations are carried out. Actually about 40 iterations are sufficient.

Exercises

1. A well-known algorithm for finding the square root of a number, a, uses the iteration $x \rightarrow f(x)$ with $f(x) = \frac{1}{2}[x + (a/x)]$. In what region is the convergence of the iteration guaranteed by the condition $f'(x) \leqslant k < 1$? Does the iteration converge for any initial value, x_0, outside this region? Are there any circumstances in which the iteration could converge to the other square root, $-\sqrt{a}$? What would happen if someone tried to use this algorithm to calculate the square root of a negative number?

2. If $f(x) = \sin x + 0.5x$, then $|f'(x)| \leqslant 0.5$ in the interval $[\pi/2, \pi]$. From this information obtain an interval that contains all points of the iteration, $x_{n+1} = f(x_n)$, with $x_0 = 2$. Estimate how many iterations will be needed to solve $2 \sin x - x = 0$ with an error less than 10^{-9}. Carry out the iteration, observe the number of iterations actually required and the magnitude of $|f'(x)|$ in the various intervals $[x_n, x_{n+1}]$.

3. Find $f'(4.5)$ (a) for $f(x) = \tan x$, (b) for $f(x) = \pi + \tan^{-1} x$.
   The equation $x = \tan x$ has a solution in the neighbourhood of $x = 4.5$. Which of the two functions mentioned above should be chosen to find this solution by means of the iteration $x_{n+1} = f(x_n)$ with $x_0 = 4.5$?

6.2. Differentiation generalized

We now consider whether the procedure of the previous section can be generalized. The first problem clearly is to produce a

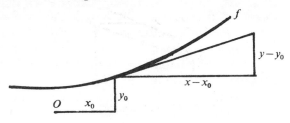

FIG. 24

definition of differentiation with much wider applications. Traditionally, differentiation enables us to find the tangent to a curve at a given point. Very often, in a diagram showing a curve with a tangent, there is a considerable stretch where the eye cannot distinguish the curve from the tangent. In other words, over a small interval the tangent gives an excellent approximation to the curve. If (x, y) is a point on the tangent to the graph of $y = f(x)$ at the point (x_0, y_0), then $y - y_0 = f'(x_0)(x - x_0)$, as in Figure 24. Now $x - x_0 \to y - y_0$ is a linear function, so the problem of differentiation is equivalent to the problem of finding a linear function, relating the change in input to the approximate change in output, the approximation being good for small changes. By now we have come to associate the term 'linear function' with a much wider class of operations than the traditional $\mathbb{R} \to \mathbb{R}$, $x \to mx$, so to obtain our generalization we need only clarify what we mean by 'a good approximation'. Here too traditional calculus suggests what should be done. For instance, when $f(x) = x^3$, $f(x_0 + h) - f(x_0) = 3hx_0^2 + 3h^2x_0 + h^3$. If h is of the order, say, of 10^{-6}, then h^2 and h^3 are of orders 10^{-12} and 10^{-18} and we may well write $f(x_0 + h) - f(x_0) = 3h_0^2 + \dots$, where the dots represent terms that are small compared to h. If we use $e(h)$ to represent the terms omitted—e for error—a formal statement is that if $f(x_0 + h) - f(x_0) = mh + e(h)$, where $e(h)/h \to 0$ as $h \to 0$, then $f'(x_0)$ exists and has the value m.

We now take a very small step in the direction of generalization, and consider functions $\mathbb{R}^2 \to \mathbb{R}^2$. Consider for example the function f that sends the vector $u = (x, y)$ to $w = (q, p)$ where $p = xy$ and $q = x^2 + y^2$. Let $h = (a, b)$. When (x, y) changes to $(x + a, y + b)$, the changes in p and q are $\Delta p = ya + xb + \dots$ and $\Delta q = 2xa + 2yb + \dots$, where the omitted terms involve a^2, ab, b^2. Thus we may write

$$\begin{pmatrix} \Delta q \\ \Delta p \end{pmatrix} \simeq \begin{pmatrix} 2x & 2y \\ y & x \end{pmatrix} \begin{pmatrix} a \\ b \end{pmatrix}. \tag{1}$$

The vector (a, b) in this equation represents h, the change in input, and the approximate change in output is given by Mh, where M is the 2×2 matrix in equation (1).

The error in our approximation is given by $e(h) = (a^2 + b^2, ab)$. We cannot keep our earlier condition, $e(h)/h \to 0$, since in general it is not possible to divide one vector by another. However we are not really interested in the direction of the vector $e(h)$, since we propose to neglect this vector. All we want to be sure of, is that its size is negligible compared with that of h. Sizes are measured by norms. So the condition, $\|e(h)\|/\|h\| \to 0$ as $\|h\| \to 0$, is meaningful and is all we need demand. We are thus led to reformulate our definition of differentiation in a way that suits the above example and works equally well for a function $f : \mathcal{X} \to \mathcal{Y}$, where $\mathcal{X}$ and $\mathcal{Y}$ are any Banach spaces:

Definition. If $f(u_0 + h) - f(u_0) = Mh + e(h)$, where M is a bounded linear operator and $\|e(h)\|/\|h\| \to 0$ as $\|h\| \to 0$, the function f is called Fréchet-differentiable at the point u_0 and we define $f'(u_0) = M$.

The term 'Fréchet-differentiable' is used because this definition is due to Maurice Fréchet, who published it in 1925. Requiring M to be bounded is much like requiring the number $m = f'(x_0)$ in traditional calculus to be finite. The convenience of this requirement can be seen from our earlier treatment of iteration, where the condition $|f'(x)| \le k$ was used. In the more general situation this is replaced by $\|f'(x)\| \le k$, an inequality which can only be fulfilled if $f'(x)$ is bounded.

Figure 25 illustrates the mapping $(x, y) \to (x^2 + y^2, xy)$ considered in our example of differentiation. It is based on the following table;—

u	(x, y)	$(q$	$, p$	$)$	w
A_0	$(1, 0\ )$	$\to (1$	$, 0$	$)$	A_1
B_0	$(1, 0.5)$	$\to (1.25,$	0.5	$)$	B_1
C_0	$(1, 1.0)$	$\to (2$	$, 1$	$)$	C_1
D_0	$(1, 1.5)$	$\to (3.25,$	1.5	$)$	D_1
E_0	$(1, 2.0)$	$\to (5$	$, 2$	$)$	E_1
F_0	$(1, 2.5)$	$\to (7.25,$	2.5	$)$	F_1

This figure helps to show why $f'(u)$ now has to be a matrix M, and not just a real number. For a change in the input such as that represented by the vector $D_0 E_0$ produces a change in the output

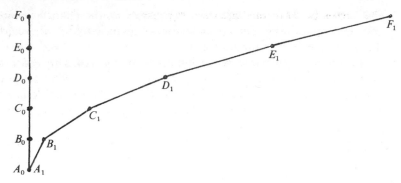

FIG. 25

given by D_1E_1. If we are to have an approximate equation $\Delta w \simeq f'(u)\,\Delta u$, then $f'(u)$ must be an operator capable of changing the vector D_0E_0 into a vector approximating D_1E_1. Multiplication by a number cannot do this, for it leaves the direction of a vector invariant. Multiplication by a matrix can change both length and direction, so it is not at all surprising that $f'(u)$ turns out to be such a transformation.

In general, since $f'(u)$ is defined by a limit with $\Delta u \to 0$, the equation $\Delta w \simeq f'(u)\,\Delta u$ will only be an approximation. However with quadratic functions a precise result is possible. In traditional calculus if $\phi(x) = ax^2 + bx + c$, then $[\phi(x+h) - \phi(x)]/h = a(2x+h) + b = \phi'(x+\tfrac{1}{2}h)$, so exactly equal to the derivative at the mid-point of the interval in question. The function f in our example is a quadratic, and the same thing happens here. If we calculate the value of f' at $(1, 1.75)$, the mid-point of D_0E_0, we get exactly the matrix required to map D_0E_0 to D_1E_1. In fact we find

$$\begin{pmatrix} 2 & 3.5 \\ 1.75 & 1 \end{pmatrix}\begin{pmatrix} 0 \\ 0.5 \end{pmatrix} = \begin{pmatrix} 1.75 \\ 0.5 \end{pmatrix}.$$

If we imagine that, as time passes, the point u moves in some continuous manner along the line A_0F_0, and w in consequence moves along a smooth curve (in fact a parabola), the vectors D_0E_0 and D_1E_1 could be used to give rather crude estimates of the velocities of u and w, on dividing them by the corresponding time Δt. With shorter intervals, better estimates would be obtained, and we might suspect that the equation $\Delta w \simeq f'(u)\,\Delta u$ would in the limit lead to the conclusion $dw/dt = f'(u)\,du/dt$. In other words, $f'(u)$ is the operator that transforms the velocity of the input to the velocity of the output. This result is in fact true for functions $f: \mathcal{X} \to \mathcal{Y}$ with arbitrary Banach spaces $\mathcal{X}$ and $\mathcal{Y}$, and is easily proved from the definition of derivative, which amounts to $\Delta w = M \Delta u + e(\Delta u)$ and so gives

$$\frac{\Delta w}{\Delta t} = M \frac{\Delta u}{\Delta t} + \frac{e(\Delta u)}{\Delta t}.$$

We are assuming that u has a velocity, that is, that $\Delta u / \Delta t \to du/dt$ as $\Delta t \to 0$. We can show that $\|e(\Delta u)/\Delta t\| \to 0$ by using the fact that $\|e(\Delta u)\|/\|\Delta u\| \to 0$ and that $\|\Delta u\|/|\Delta t|$ tends to a finite limit as $\Delta t \to 0$.

This result is very close to the original ideas of calculus. Newton, when considering $y = f(x)$, thought of x and y as physical quantities changing with velocities dx/dt and dy/dt, which he wrote as $\dot{x}$ and $\dot{y}$. The derivative, $f'(x)$, was the ratio of these velocities $\dot{y}/\dot{x}$. We cannot carry this result over to vector spaces because it involves division, but if we write the equation as $\dot{y} = f'(x)\dot{x}$, we have exactly the form of the result derived above.

Exercises

1. Find the 2×2 matrix that represents $f'(x, y)$ when $f(x, y)$ is specified by

(a) $(-y, x)$
(b) $(x + y, x - y)$
(c) $(x + y, x + y)$
(d) $(x^2 - y^2, 2xy)$
(e) $(e^x \cos y, e^x \sin y)$.

2. In section 6.2, generalized differentiation was discussed in the context of functions $\mathbb{R}^2 \to \mathbb{R}^2$. Adapt this discussion (a) to functions, $\mathbb{R}^2 \to \mathbb{R}$, such as $(x, y) \to x^2 + y^2$, (b) to functions $\mathbb{R} \to \mathbb{R}^2$ such as $t \to (a \cos t, b \sin t)$.

6.3. Generalizing the mean value theorem

Now that we have a generalized definition of differentiation, can we apply it to test the convergence of iteration along the lines discussed at the beginning of this chapter? That argument depended on a mean value theorem; if $f'(x)$ exists for $b \leq x \leq c$, then some X between b and c makes $f(c) - f(b) = f'(X)(c - b)$. As was then mentioned, this means that the tangent at some point Z is parallel to the chord BC. In Figure 23, Z is in fact the point of the arc BC that is farthest from the chord BC. For if BC has the equation $y = mx + k$, and $\phi(x) = f(x) - (mx + k)$, when $\phi(x)$ is a maximum or

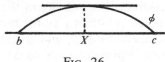

FIG. 26

minimum $0 = \phi'(x) = f'(x) - m$, so $f'(x) = m$, the gradient of *BC*. The proof of the mean value theorem therefore depends on showing that $\phi'(x) = 0$ somewhere between *B* and *C*. As the arc and the chord meet at *B* and *C*, $\phi(b) = \phi(c) = 0$ and Rolle's Theorem assures us that $\phi'(X) = 0$ for some *X* between *b* and *c*, as indicated in Figure 26, provided $\phi'(x)$ exists for *x* in $[b, c]$.

However, Rolle's Theorem has no counterpart in functional analysis. We need go no further than a function $\mathbb{R} \to \mathbb{R}^2$ to find a counterexample. Let ϕ map $t \to (x, y)$ with $x = t - t^2$, $y = t - t^5$. Then $\phi'(t) = (1 - 2t, 1 - 5t^4)$. Now $\phi(0) = (0, 0)$ and $\phi(1) = (0, 0)$, but there is no *T* with $0 < T < 1$, making $\phi'(T) = 0$. The reason for the difference is clear. When we are dealing with a single number that starts at 0 and returns to 0, there will be a moment when it stops growing and begins to decrease, or vice versa. But when the output involves two numbers, *x* and *y*, there is no such special moment. Figure 27 shows the path described by (x, y) as *t* goes from 0 to 1. The number *x* ceases growing when $t = 0.5$ but *y* continues to grow until $t = 0.67$ approximately.

Fortunately, we do not need the theorem in precisely the traditional form. We were not interested in the equation $f(c) - f(b) = f'(X)(c - a)$ for its own sake, but only as a step towards the conclusion that $|f(c) - f(b)| \leqslant k \, |c - b|$ since $|f'(X)| \leqslant k$. We are still able to reach a conclusion of this type.

Suppose, for example, we have a function *f*, $\mathbb{R}^2 \to \mathbb{R}^2$, with

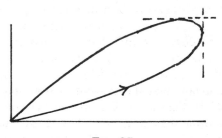

FIG. 27

$w = f(u)$ and we know that in some region $\|f'(u)\| \leq k$. This means that the matrix, $f'(u)$, when it acts on a vector, does not magnify the length of the vector more than k times. But we saw earlier that, when $w = f(u)$, and u moves about, $f'(u)$ maps the velocity du/dt to the velocity dw/dt. What our information $\|f'(u)\| \leq k$, amounts to is that the speed of w at any moment is not more than k times the speed of u. Suppose then that u moves from u_1 to u_2 by a certain path, and that the relationship $w = f(u)$ compels w to move from w_1 to w_2. If the speed at which w moves never exceeds k times the speed of u, it is reasonable to conclude that the length of the journey made by w cannot be more than k times that made by u. Now suppose u goes straight from u_1 to u_2. The distance u travels is then $\|u_2 - u_1\|$, so the length of the curve described by w cannot exceed $k \|u_2 - u_1\|$. The length of the chord from w_1 to w_2 is probably less than this, but certainly cannot be more. So we have $\|w_2 - w_1\| \leq k \|u_2 - u_1\|$.

The conclusion we have reached here is in fact true for mappings $f : \mathcal{X} \to \mathcal{Y}$ with $\mathcal{X}$ and $\mathcal{Y}$ arbitrary Banach spaces. To turn the argument above into a proof of this general theorem, we would have to explain carefully what is meant by the length of a curve in any Banach space, which might be difficult. It seems better to state the theorem we have been led to conjecture and to sketch a different proof, which itself involves an interesting and useful principle. Our theorem is:

THEOREM 1. If, at every point of the line segment joining u_1 to u_2, the derivative $f'(u)$ is defined and

$$\|f'(u)\| \leq k, \quad \text{then} \quad \|f(u_2) - f(u_1)\| \leq k \|u_2 - u_1\|.$$

We begin by considering the proof for the particular case when the spaces $\mathcal{X}$ and $\mathcal{Y}$ are both Euclidean planes, that is, for $f : \mathcal{E}^2 \to \mathcal{E}^2$. We shall then see if this proof can be generalized.

We imagine a point u starting at u_1 at time $t = 0$ and moving in a straight line at constant speed to reach u_2 at time $t = 1$. As it goes from u_1 to u_2 in unit time, its velocity, du/dt, is $u_2 - u_1$. Let $w = f(u)$. Then $dw/dt = f'(u) \, du/dt = f'(u)(u_2 - u_1)$. Now suppose co-ordinate axes brought in, with the origin at w_1, the first axis joining w_1 to w_2 and the second axis being perpendicular to it. Here w_1 and w_2 correspond to u_1 and u_2, and represent the positions of w at times $t = 0$ and $t = 1$. Let s be the first co-ordinate of w with reference to

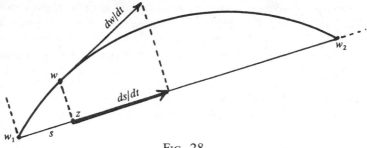

Fig. 28

these axes; the second co-ordinate does not interest us. Let z be the foot of the perpendicular from w to the first axis, as in Figure 28. As the time passes, z moves along the straight line from w_1 to w_2. Its velocity, ds/dt, is the component of dw/dt along this line, so the speed of s cannot at any time exceed the speed of w. That is $|ds/dt| \leqslant \|dw/dt\|$. Now

$$\|dw/dt\| = \|f'(u)(u_2 - u_1)\| \leqslant \|f'(u)\| \cdot \|u_2 - u_1\| \leqslant k \|u_2 - u_1\|,$$

since $\|f'(u)\| \leqslant k$ for all u on the line segment joining u_1 to u_2. Accordingly the speed of s never exceeds $k \|u_2 - u_1\|$. As all the motion occurs between $t = 0$ and $t = 1$, this same number gives an upper estimate for the distance gone, and as z goes from w_1 to w_2, this distance is $\|w_2 - w_1\|$. We now have the result we want, that $\|w_2 - w_1\| \leqslant k \|u_2 - u_1\|$.

In a formal proof, our informal argument about the maximum speed and the time taken would be replaced by the classical mean value theorem, which is equivalent to it.

The essential idea in this proof was to reduce the two-dimensional problem to a one-dimensional problem by considering only the first co-ordinate of each vector. Thus s is the first co-ordinate of $w - w_1$, ds/dt is the first co-ordinate of dw/dt, and $\|w_2 - w_1\|$ is the first co-ordinate of $w_2 - w_1$. Let L stand for this operation of replacing every vector by its first co-ordinate. We note the following facts about L;—

(i) L is a linear operator.
(ii) $\|L\| = 1$, for the magnitude of the first co-ordinate is never greater than the length of the vector; it equals that length if the vector lies along the first axis.
(iii) $L(w_2 - w_1) = \|w_2 - w_1\|$.

(iv) L behaves like a constant in relation to the operation of differentiation, d/dt. For if $v = (x, y)$ in the axes introduced above, $dv/dt = (dx/dt, dy/dt)$. $Lv = x$ and $L(dv/dt) = dx/dt$, by taking the first co-ordinates. It follows that $(d/dt)Lv = (d/dt)x = dx/dt = L(dv/dt)$.

Property (iv) in fact is a consequence of property (i). As we shall see in section 7.2 for any linear operator L, $(Lf)' = Lf'$.

These observations provide us with a means of generalizing our proof, for it can be shown that the proof depends only on the existence of an operator with these properties, and that such an operator exists in the general case.

First, then, we must check that only these properties were used in our proof above, and that we were not taking advantage of special circumstances for Euclidean planes. If the argument above is examined, it will be found that the result $dw/dt = f'(u)(u_2 - u_1)$ was reached by arguments of a very general nature. Suppose then at this point we assume the existence of an operator $L : \mathcal{Y} \to \mathbb{R}$ with the four properties, and define $s = L(w - w_1)$. Then $ds/dt = L\, d/dt(w - w_1)$ by property (iv), so $ds/dt = L\, dw/dt$ since w_1 is constant. As $\|L\| = 1$, $\lceil ds/dt \rceil \leqslant \|dw/dt\|$. (For a vector in one dimension, represented by a single real number, the norm is given by the absolute value of that number. This is why there is a single upright line, and not a double one, on each side of the symbol ds/dt.) Then, as we had earlier,

$$\|dw/dt\| = \|f'(u)(u_2 - u_1)\| \leqslant \|f'(u)\| \cdot \|(u_2 - u_1)\| \leqslant k \|u_2 - u_1\|.$$

We now return to our definition, that $s = L(w - w_1)$. At $t = 0$, $w = w_1$, so $s(0) = L0 = 0$. At $t = 1$, $s(1) = L(w_2 - w_1) = \|w_2 - w_1\|$ by property (iii). Now using the classical calculus result (mean value theorem) that $|s(1) - s(0)|$ cannot exceed the maximum value of $|ds/dt|$, we have what we require, $\|w_2 - w_1\| \leqslant k \|u_2 - u_1\|$.

We may feel sceptical about the existence of such an operator L for general Banach spaces, since in our special case, L was constructed by means of dropping perpendiculars. In most vector spaces the word 'perpendicular' is totally meaningless. But it should be noticed that this word does not occur in any of the property statements (i) to (iv). In fact there is an important general theorem, which we shall discuss in section 10.2, guaranteeing the existence of a suitable $L : \mathcal{Y} \to \mathbb{R}$. An operator with $\mathbb{R}$ (or $\mathbb{C}$) as the output space is called a *functional*, for historical reasons. The theorem in question thus takes the form;—

THEOREM 2.   If a is a vector in any Banach space, $\mathscr{S}$, there exists a bounded, linear functional $L : \mathscr{S} \to \mathbb{R}$, such that $La = \|a\|$ and $\|L\| = 1$.

If we choose $w_2 - w_1$ for a, this theorem ensures the existence of an operator L with properties (i), (ii), (iii). Property (iv), as has already been mentioned, is a consequence of property (i). If we accept this, and the truth of Theorem 2, then Theorem 1 is established for any function f involving Banach spaces.

The inequality in Theorem 1 is what we need to establish that f is a contraction mapping when $\|f'(u)\| \leqslant k < 1$. Theorem 1 depends on this estimate for $f'(u)$ holding at every point of the line segment joining u_1 to u_2. We can be sure that this will be so if u_1 and u_2 lie in a convex region, for all points of which $\|f'(u)\| \leqslant k$. Our conclusion is summed up in the following theorem:

THEOREM 3.   Let $\mathscr{S}$ be a Banach space and f a function $\mathscr{S} \to \mathscr{S}$. In some convex region of $\mathscr{S}$, let the derivative f' be defined, with $\|f'(u)\| \leqslant k$ for each point u of the region. Then for any two points, u_1, u_2 of the region $\|f(u_2) - f(u_1)\| \leqslant k \|u_2 - u_1\|$.

If a sequence is formed by the iteration $v_{n+1} = f(v_n)$, and we wish to apply Theorem 3 to establish its convergence, we must make sure that the points of the sequence lie in the specified region. If they do, the first link of the chain will have length $\|v_1 - v_0\|$ and the total length of the chain will not exceed $\|v_1 - v_0\|(1 + k + k^2 + \dots) = \|v_1 - v_0\|/(1 - k) = r$ say. If the ball $\bar{B}(v_0, r)$ is contained in the region, we can be sure the chain will stay in the region with $\|f'(u)\| \leqslant k$ since the chain is contained in the ball $\bar{B}(v_0, r)$.

This argument may appear circular. We find r, the radius of the ball, by supposing the chain stays in the region. We then use r to show that the chain does stay in the region. However the reasoning is in fact valid, and can be established step by step. We suppose r has been calculated and it has been checked that $\bar{B}(v_0, r)$ does lie in the region. Now $\|v_1 - v_0\| < r$, so v_1 is in the ball and hence the line segment joining v_0 to v_1 has $\|f'(v)\| \leqslant k$ at each of its points. Accordingly

$$\|v_2 - v_1\| = \|f(v_1) - f(v_0)\| \leqslant k \|v_1 - v_0\|$$

by Theorem 1. By the triangle inequality $\|v_2 - v_0\| \leqslant \|v_1 - v_0\| (1 + k)$, the combined length of the first two links of the chain. This means that v_2 is in $\bar{B}(v_0, r)$, so the line segment v_1 to v_2 is also, and we can deduce $\|v_3 - v_2\| \leqslant k \|v_2 - v_1\|$. The argument continues; a formal proof would use the polygon inequality and mathematical induction.

We sum this up in the form of a theorem;

THEOREM 4. If the closed ball $\bar{B}(v_0, r)$ is contained in a region where $\|f'(v)\| \leqslant k < 1$, where $r = \|f(v_0) - v_0\|/(1 - k)$, then the iteration $v_{n+1} = f(v_n)$, starting from v_0, will converge to a point in this ball.

In section 5.7 we observed that the convergence of the iteration $v_{n+1} = f(v_n)$ did not always give a solution of $v = f(v)$. The proof that a solution would be reached depended on the assumption of f being continuous. In the present context this does not cause any trouble. In functional analysis, as in traditional calculus, a differentiable function is of necessity continuous. It follows that the limit point of the iteration specified in Theorem 4 will in fact satisfy the equation $v = f(v)$.

6.4   Some worked examples

In section 2.1, the fourth example considered the iteration $(x, y) \rightarrow (xy + x + 0.07, x^2 + y^2 + y - 0.41)$ as a way of solving the equation system $xy = -0.07$, $x^2 + y^2 = 0.41$. Clearly this was not intended as a practical recommendation for computing, since x is a root of the equation $x^4 - 0.41x^2 + 0.0049 = 0$ and can be found by two extractions of square roots. We are considering it only as a very simple situation in which the meaning of the derivative f' can be seen. We ignore for this purpose the algebraic simplicity of the problem and the high degree of symmetry shown by the intersections of the circle and the rectangular hyperbola.

There is a standard formula that gives the derivative of a function $f: \mathbb{R}^2 \rightarrow \mathbb{R}^2$, $(x, y) \rightarrow (p, q)$. If a change $(\Delta x, \Delta y)$ in the input produces a change $(\Delta p, \Delta q)$ in the output, and the function is differentiable, we have

$$\Delta p = \frac{\partial p}{\partial x} \Delta x + \frac{\partial p}{\partial y} \Delta y + \ldots$$

$$\Delta q = \frac{\partial q}{\partial x} \Delta x + \frac{\partial q}{\partial y} \Delta y + \ldots.$$

Thus

$$f'\begin{pmatrix} x \\ y \end{pmatrix} = \begin{pmatrix} \dfrac{\partial p}{\partial x} & \dfrac{\partial p}{\partial y} \\[2mm] \dfrac{\partial q}{\partial x} & \dfrac{\partial q}{\partial y} \end{pmatrix}.$$

For the function defined by our iteration

$$f'\begin{pmatrix} x \\ y \end{pmatrix} = \begin{pmatrix} y+1 & x \\ 2x & 2y+1 \end{pmatrix}.$$

The ℓ_∞ norm of this is given by the larger of the two numbers $|y+1|+|x|$, $|2x|+|2y+1|$. A preliminary survey of the equation system would indicate that solutions might exist in the neighbourhood of the points $(0.1, -0.6)$, $(0.6, -0.1)$, $(-0.6, 0.1)$ and $(-0.1, 0.6)$. At these four points $\|f'\|_\infty$ has the values 0.5, 2, 2.4, 2.4. Only the first of these is less than 1. This shows that the solution near $(0.1, -0.6)$ is likely to be an attractive fixed point and strongly suggests that the other three are repulsive. It does not prove that they are repulsive since, as the matrix M in section 3.5 illustrated, it is possible to have $\|M\|_\infty > 1$ and yet $\sum M^n v$ convergent. In fact, in this very simple example, it can be proved by considering eigenvalues that all three points are indeed repulsive. If we take (x_0, y_0) as $(0.6, -0.1)$, or another of these three points, all that happens in the first dozen or so iterations is that the point wanders across to the neighbourhood of $(0.1, -0.6)$.

The value 0.5 for f' at $(0.1, -0.6)$ suggests that we may be able to prove convergence by considering a value of k in Theorem 4 that is somewhat larger than 0.5. It will be found that $k = 0.7$ works with a certain margin, and $k = 0.6$ works with nothing to spare. In Figure 29, the square $ABCD$ contains the points for which $|y+1|+|x| \leqslant 0.6$ and $EFGH$ those for which $|2x|+|2y+1| \leqslant 0.6$. The condition $\|f'(x, y)\| \leqslant 0.6$ holds in the region common to these two squares. If $v_0 = (0.1, -0.6)$, $v_1 = (0.11, -0.64)$ and $\|v_1 - v_0\| = 0.04$. With $k = 0.6$, $\|v_1 - v_0\|/(1-k)$ is 0.1. The closed ball $B(v_0, 0.1)$ is shown shaded. It just manages to fit in the region where $\|f'(v)\| \leqslant 0.6$, so the conditions of Theorem 4 apply. We can be sure the iteration will converge and have its limit point in the shaded region.

Now of course there are an infinity of iterations that correspond to a given system of equations. It would be perfectly possible to

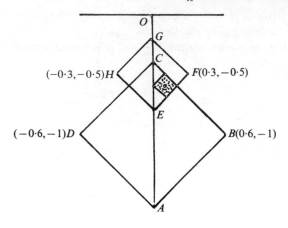

FIG. 29

consider $f:(x, y) \to [-0.07/y, \sqrt{(0.41 - x^2)}]$. This has

$$f'\begin{pmatrix} x \\ y \end{pmatrix} = \begin{pmatrix} 0 & 0.07/y^2 \\ \dfrac{-x}{\sqrt{(0.41 - x^2)}} & 0 \end{pmatrix}$$

Then $\|f'\|_\infty$ takes the value 0.19 at $(0.1, -0.6)$ and $(-0.1, 0.6)$, and takes the value 7 at $(0.6, -0.1)$ and $(-0.6, 0.1)$. The convention that the square root is always taken positive means that we can only hope to reach solutions with y positive by this method, and it appears that the solution near $(-0.1, 0.6)$ is likely to be the only attractive one of these. The value 0.19 suggests rapid convergence in its neighbourhood.

The condition $\|f'(v)\|_\infty \leqslant k$ leads to the inequalities $|x| \leqslant k\sqrt{[0.41/(1 + k^2)]}$, $|y| \geqslant \sqrt{(0.07/k)}$. If we try the value $k = 0.2$ we find it is unsatisfactory, as the initial point $v_0 = (-0.1, 0.6)$ lies too close to the boundary of the corresponding region, but $k = 0.3$ works well; $\|f'(v)\|_\infty \leqslant 0.3$ for $|x| \leqslant 0.1840$, $|y| \geqslant 0.4830$.

As $v_1 = (-0.1167, 0.6323)$, $\|v_1 - v_0\| = 0.0323$ and $\|v_1 - v_0\|/(1 - 0.3) = 0.0461$. Both co-ordinates can vary by this amount from $(-0.1, 0.6)$ while remaining in the region for $k = 0.3$. Thus the convergence of the iteration is assured.

Comparing this method of iteration with the earlier one, we

notice that the region in which $\|f'(v)\| \leqslant k$ is much larger and the value of k is noticeably smaller.

The sixth example in section 2.1 was concerned with the iteration $f_n \rightarrow f_{n+1}$ where

$$f_{n+1}(x) = x + \int_0^x [f_n(t)]^2 \, dt.$$

We assume $x \geqslant 0$.

Let Φ denote the function $f \rightarrow g$, where

$$g(x) = x + \int_0^x [f(t)]^2 \, dt.$$

We require Φ'. If f changes to $f + h$, then $[f(t)]^2$ becomes $[f(t)]^2 + 2f(t)h(t) + [h(t)]^2$. Just as in the traditional differentiation of x^2, the h^2 term makes no contribution to the final result. We may write $\Delta g \cdot (x) = \int_0^x 2f(t)h(t) \, dt + \ldots$ and find

$$\Phi'(f)h \cdot (x) = \int_0^x 2f(t)h(t) \, dt.$$

The full stop is placed in $\Delta g \cdot (x)$ in the hope of emphasizing a point which, if unnoticed, would lead to confusion. In elementary calculus we meet expressions such as $\Delta g(x) = g'(x) \Delta x + \ldots$ and understand that we have an approximate expression for the change in the value of one and the same function g when a change Δx takes place in x. The situation with Φ' is entirely different. We suppose f changes to a different function $f + h$ as a result of which g changes to a new function $g + \Delta g$. We are interested in the value taken by Δg at a point x which we imagine as unchanging, since nowhere in the argument does Δx appear. In the same way, as $g = \Phi f$, we could write $g(x)$ as $\Phi f \cdot (x)$, which must not be thought of as $\Phi[f(x)]$, for in modern notation $f(x)$ is a number, while Φ corresponds to a computer program that demands a function as input; it does not know what to do with a single number, $f(x)$. Indeed the reason for bringing the new symbol g into the discussion, instead of using Φf, is to avoid any misunderstanding of the symbol $\Phi f \cdot (x)$. This device, a special symbol for the output function, will be met from time to time both in this book and in the literature generally.

Our concern is with $\|\Phi'\|$. Naturally we suppose f and g to be continuous functions, so Φ is a function $\mathscr{C}[a, b] \to \mathscr{C}[c, d]$ for some suitable intervals $[a, b]$ and $[c, d]$.

Under $\Phi'(f)$, $h \to k$ where

$$k(x) = \int_0^x 2f(t)h(t) \, dt.$$

What is the largest value $\|k\|$ can have if $\|h\| = 1$? As $\|h\| = 1 \Rightarrow |h(t)| \leqslant 1$, it is clear that $|k(x)|$ cannot exceed $2x \sup\{|f(t)|\}$ with $0 \leqslant t \leqslant x$.

If we iterate, beginning with $f_0 = 0$, we find $f_n(x) \geqslant 0$ and $f_n'(x) \geqslant 0$ for all n. So, for the functions we are concerned with, $|f(t)| = f(t)$ and the maximum value is taken at x. Thus $|k(x)| \leqslant 2xf(x)$. If x is in the interval $[0, a]$, where $a > 0$, since at the outset we confined ourselves to $x \geqslant 0$, then $|k(x)| \leqslant 2af(a)$ and so $\|k\| \leqslant 2af(a)$. This means that $\|\Phi'(f)\| \leqslant 2af(a)$, since by definition $\|\Phi'(f)\|$ is the supremum of $\|k\|$ subject to $\|h\| = 1$.

We are thus assured of a contraction mapping if we keep x in an interval $[0, a]$ such that $2af(a) < 1$. Of course $f(a)$ here is a number that becomes known only in the course of the iteration. Actually there are ways of determining a value of a that certainly satisfies this condition, that can be done before the iteration is carried out. We defer this question until section 7.3, in order not to be distracted from our present theme, the nature and use of calculus operations in functional analysis.

One of the first breakthroughs in the application of functional analysis to computing was Kantorovich's generalization, in 1948, of the Newton–Raphson method for solution of equations. The essential idea of this method, in its classical form, was to approximate to a root of the equation $f(x) = 0$ by considering where a tangent to the graph of $y = f(x)$ crossed the x-axis. Here again, as in section 6.2, the central idea is that a tangent gives a linear approximation to a curve. The method uses iteration. If the tangent at x_0 crosses the axis at x_1, we find x_2 where the tangent at x_1 crosses the axis, and so

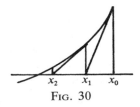

FIG. 30

on. In favourable circumstances, as illustrated in Figure 30, the iteration converges and gives a solution of the equation $f(x) = 0$.

Kantorovich developed a theory which not only generalized the Newton–Raphson method but also gave a beautiful test for its convergence and rate of convergence. He constructs a certain iteration of traditional type, that is to say, one involving a function $\mathbb{R} \to \mathbb{R}$, and shows that the generalized iteration behaves as well as, or better than, this iteration.

To justify Kantorovich's method, it is necessary to use the idea of integration in Banach spaces, so before dealing with his work we take a quick glance at generalized integration.

Exercises

1. Let F be defined by $f \to g$. For a number of cases the value of $g(x)$ is shown below. For each of these, write the expression $F'(f)h \cdot (x)$ which defines the derivative $F'(f)$.

(a) $[f(x)]^2$.

(b) $[f(x)]^n$.

(c) $x^3[f(x)]^2$.

(d) $\sin f(x)$.

(e) $\displaystyle\int_0^x t[f(t)]^2 \, dt$.

(f) $f'(x)$.

(g) $f(x)f'(x)$.

(h) $\displaystyle x\int_0^1 [f'(t)]^2 \, dt$.

(i) $\displaystyle\int_0^x [f(t)]^2 + [f'(t)]^2 \, dt$.

(j) $\displaystyle\left[\int_0^x f(t) \, dt\right]^2$.

2. Let S be the function $\mathscr{C}[0, 1] \to \mathscr{C}[0, 1]$, $f \to g$, where

$$g(x) = (x/8) + \int_0^x [f(t)]^2 \, dt.$$

Find the derivative $S'(f)$ and show that $\|f\| \leq 0.25 \Rightarrow \|S'(f)\| \leq 0.5$. Hence show that the iteration, $f_{n+1} = Sf_n$ with $f_0 = 0$, converges to a function belonging to $\bar{B}(0, 0.25)$.

6.5. The idea of an integral

In the classical theory of integration, the integral is defined as the limit of a sum. To define $\int_a^b f(x) \, dx$ we break $[a, b]$ up into a number of shorter intervals and consider $\sum f(x) \, \Delta x$, where x represents some value chosen in an interval of length Δx. If f is continuous on the finite interval $[a, b]$, it is shown that $\sum f(x) \, \Delta x$ tends to a limit, regardless of how the values x are chosen inside the intervals, when the number of intervals tends to infinity and the length of the longest interval tends to nought.

What form should a generalization of this take? The number Δx is a difference, $x_{n+1} - x_n$, so we must generalize to a system in which subtraction is possible. This suggests that we replace x by v, an element of a vector space. Now $\sum f(v)\,\Delta v$ is to be meaningful, so the things $f(v)\,\Delta v$ must be capable of being added; this suggests that they too should be elements of a vector space. What then should $f(v)$ be? It will not do to have $f(v)$ a member of the same vector space as Δv, for as a rule there is no way of multiplying two vectors in a given space. Of course $f(v)$ could be simply a real number, but this would hardly be a generalization. We get a hint from our earlier work on differentiation. There we had equations of the type $w = M\,\Delta v + \ldots$, where M was specified by a matrix or, more generally, was a linear operator. As integration is more or less the inverse of differentiation, it is suggested that we take $f(v)$ to be a linear operator.

Consider then a very simple example. Suppose $v = (x, y)$ and

$$f(v) = \begin{pmatrix} x & y \\ 1 & 3+2y \end{pmatrix}.$$

Let $(\Delta X, \Delta Y) = f(v)\,\Delta v$.
Then

$$\begin{pmatrix} \Delta X \\ \Delta Y \end{pmatrix} = \begin{pmatrix} x & y \\ 1 & 3+2y \end{pmatrix}\begin{pmatrix} \Delta x \\ \Delta y \end{pmatrix}. \tag{2}$$

As the path of integration we take the line segment from $(0, 0)$ to $(7, 1)$. As a crude approximation to integration, we divide this path into ten equal pieces, and in $f(v)$ take v as the mid-point of the piece in question. All these intervals being equal, $\Delta x = 0.7$ and $\Delta y = 0.1$ for each interval. The mid point of the first interval is $(0.35, 0.05)$, so the contribution to the sum $f(v)\,\Delta v$ from the first interval is

$$\begin{pmatrix} 0.35 & 0.05 \\ 1 & 3.1 \end{pmatrix}\begin{pmatrix} 0.7 \\ 0.1 \end{pmatrix} = \begin{pmatrix} 0.25 \\ 1.01 \end{pmatrix}.$$

In the same way we obtain a vector $f(v)\,\Delta v$ from each of the other nine intervals, and have the following table for the ten contributions.

ΔX	ΔY	X	Y
0.25	1.01	0.25	1.01
0.75	1.03	1.00	2.04
1.25	1.05	2.25	3.09
1.75	1.07	4.00	4.16
2.25	1.09	6.25	5.25
2.75	1.11	9.00	6.36
3.25	1.13	12.25	7.49
3.75	1.15	16.00	8.64
4.25	1.17	20.25	9.81
4.75	1.19	25.00	11.00

The first two columns, headed ΔX and ΔY, show the individual contributions. The last two columns, headed X and Y, show how the co-ordinates of $\sum f(v)\,\Delta v$ grow as the summation proceeds. The final (X, Y) entry, $(25, 11)$ gives an estimate for the value of $\int f(v)\,dv$, taken by the direct path from $(0, 0)$ to (7.1). Figure 31 illustrates the data in this table. The vectors *OA*, *AB*, *BC* and so on represent the contributions of the individual intervals. The sum of these ten vectors brings us to *J* with co-ordinates $(25, 11)$.

If we take smaller and smaller intervals on the path from $(0, 0)$ to $(7, 1)$, instead of the broken line *OABCDEFGHIJ* we shall get a broken line that very closely resembles a smooth curve. This suggests that we do not need to go back to the basic definition of an

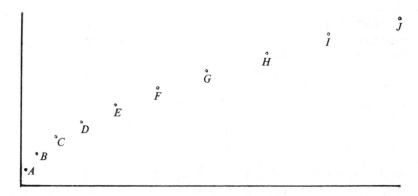

Fig. 31

integral as the limit of a sum; instead we might be able to apply the developed machinery of calculus to this problem in two dimensions. If we write $x = 7t$, $y = t$ then as the time t goes from 0 to 1, the point (x, y) will travel from $(0, 0)$ to $(7, 1)$. From equation (2), $\Delta X = x\,\Delta x + y\,\Delta y$. By substituting in this we find $\Delta X = 50t\,\Delta t$, so $X = \sum 50t\,\Delta t$, which is an approximation to $\int_0^1 50t\,dt$. In fact, if we go back to equation (2) and observe the expressions for X and Y that come from it, we can see that in the limit we are going to have

$$X = \int x\,dx + y\,dy, \quad Y = \int dx + (3 + 2y)\,dy.$$

The substitution $x = 7t$, $y = t$ then gives $X = \int_0^1 50t\,dt = 25$, $Y = \int_0^1 (10 + 2t)\,dt = 11$. These values agree exactly with our estimates by the crude arithmetical method. This agreement is owing to the fact that, in our example, the matrix involves only expressions in the first degree in x and y, for which the mid-point rule gives the exact answer. In general, we do not expect this to happen. Rather, as we divide the path of integration into ever smaller parts, we expect the sum $f(v)\,\Delta v$ to move towards some limiting position—provided, of course, that f is an integrable function. In functional analysis, as in classical, it can be proved that, for a sufficiently simple path, this limit will exist if f is continuous. It may of course exist in other circumstances.

The device used above, by which an integral along a line (or curve), was changed into an integral with respect to the time t, is not restricted to finite dimensional problems. If a vector v can be made to travel along a prescribed path by taking $v = v(t)$, with t going from a to b, and if the derivative $v'(t)$ exists and is continuous for these values of t, it can be proved that

$$\int f(v)\,dv = \int_a^b f(v)v'(t)\,dt.$$

In our later work on the Newton–Raphson method, we shall need this result only in the simplest case, when v travels along the line segment from $v(a)$ to $v(b)$. We may suppose this to happen with constant velocity; then $v'(t) = [v(b) - v(a)]/[b - a]$.

Another familiar result survives in that simple situation, namely

$$\int \phi'(v)\,dv = \phi(v_1) - \phi(v_0),$$

the integral in this case being along the line joining v_0 to v_1. It is assumed that $\phi'(v)$ is defined for each v on this line segment, and that ϕ and ϕ' are continuous.

Of familiar type also is the result that, if $v(t)$ depends continuously on the time, and

$$\phi(T) = \int_0^T v(t)\, dt,$$

then $\phi'(T) = v(T)$.

If we avoid analytical subtleties by confining our attention to situations in which all functions and their derivatives are continuous, integration can be pictured in a simple way. We visualize $\int_a^b v(t)\, dt$ as the change in position between times $t = a$ and $t = b$ of a point moving with velocity $v(t)$. If we put an expression $\int f(w)\, dw$ in the form $\int f(w)w'(t)\, dt$ we can see this as related to a point whose velocity is given by $f(w)w'(t)$.

There are three inequalities frequently used that can be thought out and remembered easily with the help of this picture of integration.

1. If a point moves with velocity $v(t)$ between $t = a$ and some later instant $t = b$, the distance of its final from its initial position is given by $\left\| \int_a^b v(t)\, dt \right\|$. Suppose now we wish to make this distance as large as possible, subject to the speed $\|v(t)\|$ being prescribed at each instant. That is to say, we control the direction of movement but not the magnitude of the velocity—our hands are on the steering wheel but the accelerator has escaped from our control. Clearly we shall get farthest by travelling always in the same direction. The distance covered will be $\int_a^b \|v(t)\|\, dt$. The inequality, expressing the fact that this distance cannot be exceeded is

$$\left\| \int_a^b v(t)\, dt \right\| \leq \int_a^b \|v(t)\|\, dt.$$

2. If we apply inequality (1) to $\int_a^b f(w)w'(t)\, dt$ we get

$$\left\| \int_a^b f(w)w'(t)\, dt \right\| \leq \int_a^b \|f(w)w'(t)\|\, dt.$$

Considering now the operator $f(w)$ acting on the vector $w'(t)$, by the definition of operator norm we have

$$\|f(w)w'(t)\| \leq \|f(w)\|\, \|w'(t)\|.$$

So

$$\left\| \int_a^b f(w)w'(t) \, dt \right\| \leq \int_a^b \|f(w)\| \, \|w'(t)\| \, dt.$$

3. Finally, we know

$$\phi(w_1) - \phi(w_0) = \int \phi'(w) \, dw = \int_a^b \phi'(w)w'(t) \, dt,$$

where $w_0 = w(a)$ and $w_1 = w(b)$. Applying inequality (2) with $f(w) = \phi'(w)$ we have

$$\|\phi(w_1) - \phi(w_0)\| \leq \int_a^b \|\phi'(w)\| \, \|w'(t)\| \, dt.$$

Here, of course, as in (1) and (2), it is assumed that $b > a$.

6.6. The Newton–Raphson method

The well known method for solving equations, developed in the seventeenth century by Newton and Raphson, was mentioned at the end of section 6.4 and illustrated in Figure 30. The tangent to $y = f(x)$ at the point x_0 has the equation $y = f(x_0) + (x - x_0)f'(x_0)$. If instead of finding the intersection of the curve $y = f(x)$ with the x-axis, we find that of the tangent, we are led to $x_1 = x_0 - f(x_0)/f'(x_0)$.

The method generalizes immediately to Banach spaces. If X is any Banach space, and the function $f; \mathcal{X} \to \mathcal{X}$ is differentiable, we have

$$f(x_0 + h) = f(x_0) + f'(x_0)h + \ldots$$

and we obtain a linear approximation, the analogue of the tangent, by neglecting the part represented by dots. We have to be careful about the order of products. For instance, in a finite number of dimensions, $f'(x_0)$ is represented by a matrix acting on the vector h. By the convention we are using the matrix, $f'(x_0)$, must be written in front of the vector, h. Equating the approximate value of $f(x_0 + h)$ to 0, we obtain $h = -[f'(x_0)]^{-1}f(x_0)$ and so

$$x_1 = x_0 + h = x_0 - [f'(x_0)]^{-1}f(x_0). \tag{3}$$

Even in the classical case, $f: \mathbb{R} \to \mathbb{R}$, we are let in for a certain amount of work if we simply iterate the function defined by (3), for

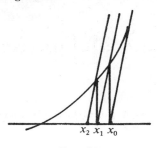

$$x_2 \ x_1 \ x_0$$

FIG. 32

as we go from x_n to x_{n+1} we must compute $f'(x_n)$ and divide by it. The work is very much increased in the new situation. Even for $f: \mathbb{R}^n \to \mathbb{R}^n$, $f'(x_n)$ is specified by an $n \times n$ matrix, which we have to compute for each new x_n, and still worse, find its inverse. For $f: \mathscr{C}[a, b] \to \mathscr{C}[a, b]$, we may have to find the inverse of an integral operator.

For this reason, a modified procedure is often used. In the classical case, instead of the tangents at the successive points corresponding to x_n, we use lines parallel to the initial tangent. (See Figure 32.) In the generalized situation, a similar procedure leads to the equation

$$x_{n+1} = x_n - [f'(x_0)]^{-1} f(x_n). \tag{4}$$

Now, just the one inverse, $[f'(x_0)]^{-1}$, is calculated and is used throughout.

We will illustrate this method by applying it to the differential equation

$$y' + y^2 = 1/(x + 1)^2. \tag{5}$$

The initial condition $y(0) = 1$ is given, and we shall be interested only in $x \geq 0$.

A sequence of approximations $y_0, y_1, y_2, \dots$ will be obtained. These correspond to the sequence $x_0, x_1, x_2, \dots$ in equation (4). The variable x in equation (5) of course has no relationship at all to the x_n in equation (4).

If the right-hand side of equation (5) had been 0 instead of $1/(x + 1)^2$, the solution would have been $y_0(x) = 1/(x + 1)$. We take y_0 as our initial approximation.

If we substitute any differentiable function y in equation (5), the two sides of the equation will differ by an error

$$e(x) = y'(x) + [y(x)]^2 - 1/(1+x)^2 \qquad (6)$$

Our aim is to find a function y that makes this error nought for all $x \geq 0$. So let $f(y) = e$. We want to solve $f(y) = 0$. The modified Newton–Raphson method will be used. The work falls into several parts. In the account below, as well as solving the problem we give an explanation and justification of Kantorovich's test. In actual practice the work would be shorter, since the test would simply be quoted and applied.

1.   Our first aim is to find $f'(y)$. Suppose that, as a result of y changing to $y + h$, the error $e(x)$ becomes $e(x) + k(x) + \dots$. As $y(0) = 1$ is required we must also have $y(0) + h(0) = 1$, that is $h(0) = 0$. Of course h must be a differentiable function. We then find $k(x) = h'(x) + 2y(x)h(x)$, and $f'(y)$ is the function $h \to k$. In particular, as $y_0(x) = 1/(x+1)$, we have $f'(y_0)$ as $h \to k$ where

$$k(x) = h'(x) + [2h(x)/(x+1)]. \qquad (7)$$

2.   We now need the inverse, $[f'(y_0)]^{-1}$. If equation (7) is multiplied by $(x+1)^2$, the righthand side becomes an exact differential and can be explicitly integrated. Take the integral from 0 to x, and use the fact that $h(0) = 0$. This gives

$$(x+1)^2 h(x) = \int_0^x (s+1)^2 k(s)\, ds. \qquad (8)$$

Dividing equation (8) by $(x+1)^2$ gives $h(x)$ explicitly in terms of k. The function $k \to h$ is the inverse function we are seeking.

3.   We now come to the iteration

$$y_{n+1} = y_n - [f'(y_0)]^{-1} f(y_n). \qquad (9)$$

From equation (8) we see that $[f'(y_0)]^{-1}$ is the operator $(x+1)^{-2}\int_0^x (s+1)^2[\ \ ]\, ds$. The expression for the input function has to be inserted in the space between the square brackets. In equation (9) the input is $f(y_n)$. As $f(y) = e$, we go back to equation (6). The expression we need is the right-hand side of that equation, with the subscript n wherever y or y' occurs. This expression in fact shows the discrepancy in equation (5) corresponding to the approximate solution, y_n. This is the usual procedure in Newton–Raphson methods; the operator $[f'(y_0)]^{-1}$ acts on the discrepancy in the latest approximation. Accordingly, entering the specified expression in the square brackets, we obtain

$$y_{n+1}(x) = y_n(x) - (x+1)^{-2} \int_0^x (s+1)^2 [y_n'(s) + y_n(s)^2 - (s+1)^{-2}] \, ds.$$

Integration by parts can now be applied to the term $(s+1)^2 y_n'(s)$. When this is done, very considerable simplification takes place and the equation eventually reaches the form

$$y_{n+1}(x) = (x+1)^{-1} + (x+1)^{-2} \int_0^x 2(s+1) y_n(s)$$

$$- (s+1)^2 [y_n(s)]^2 \cdot ds. \quad (10)$$

Thus we have the iteration formula showing how to obtain y_{n+1} from y_n. Let $y_{n+1} = S(y_n)$ stand for this formula. The symbol S is that used by Kantorovich in his condition for the validity of the iteration.

4. The iteration leads to the following results.

$$y_0(x) = 1/(x+1), \quad (11)$$

$$y_1(x) = \frac{1}{x+1} + \frac{x}{(x+1)^2}, \quad (12)$$

$$y_2(x) = \frac{1}{x+1} + \frac{2\ln(x+1)}{(x+1)^2} - \frac{x}{(x+1)^3}. \quad (13)$$

After this the analytic expressions become increasingly complicated, owing to the appearance of the logarithm in y_2, and it would be convenient to use numerical integration from here on. However the above work will be sufficient for our present aim, to illustrate the method and the Kantorovich test for convergence.

5. The idea of the test is the following. We consider an iteration $\mathbb{R} \to \mathbb{R}$, $t \to \phi(t)$, $t_0 = 0$ with the following properties; (a) the first jump for y is not longer than the first jump for t, (b) S does not vary more quickly than ϕ. It then turns out that the iteration with $y_{n+1} = S(y_n)$ is at least as well-behaved as the iteration with $t_{n+1} = \phi(t_n)$, in the sense that $\|y_n - y_{n-1}\|$, the length of the nth link in the chain of y-points, never exceeds $|t_n - t_{n-1}|$, the length of the corresponding link in the chain of real numbers, t.

Naturally we have to clarify what we mean by the conditions (a) and (b). Condition (a) compares the lengths of the first links of the two chains and requires $\|y_1 - y_0\| \leqslant |t_1 - t_0| = |t_1|$ since $t_0 = 0$. Actually the numbers 0, t_1, t_2, ... steadily increase so that we can dispense with the absolute value sign in $|t_n - t_{n-1}|$. Condition (b) is to be understood as meaning that $\|S'(y)\| \leqslant \phi'(t)$ so long as $\|y - y_0\| \leqslant t$.

The comparison with the iteration $t_n \to \phi(t_n)$ will only be helpful if that iteration converges. Since ϕ' is supposed to exist, ϕ must be continuous, so the iteration will converge to a solution of $t = \phi(t)$. Call this solution T. We suppose ϕ defined in an interval that includes $[0, T]$. As we plan to have $\|S'(y)\| \leqslant \phi'(t)$ we certainly must take $\phi'(t) \geqslant 0$, so the conditions compel us to seek for an increasing function ϕ. Figure 33 shows a typical situation, the graph of $z = \phi(t)$ crossing the graph of $z = t$ where $t = T$. It is clear that the numbers 0, t_1, t_2, ... form an increasing sequence and that they converge to T.

6.   How is the statement to be justified that no link in the chain y_0, y_1, y_2, ... is longer than the corresponding link in the chain 0, t_1, t_2, ...? Assumption (a) states that the first link of the y-chain is not longer than the first link of the t-chain, so we have nothing to prove

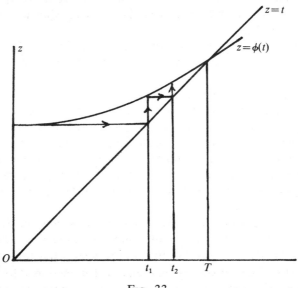

Fig. 33

about these first links. For the second link we want to prove $\|y_2 - y_1\| \leqslant |t_2 - t_1|$. As $y_2 = S(y_1)$ and $y_1 = S(y_0)$, we have $\|y_2 - y_1\| = \|S(y_1) - S(y_0)\|$, which can be estimated in terms of $\|S'(y)\|$. Let us arrange for a point to start at y_0 at time $t = 0$ and travel with constant velocity to arrive at y_1 at $t = t_1$. Its velocity will be $y'(t) = (y_1 - y_0)/t_1$. By assumption (a), $\|y_1 - y_0\| \leqslant t_1$. This shows $\|y'(t)\| \leqslant 1$; the speed of the moving point never exceeds 1. Accordingly, at any time t in the interval $[0, t_1]$, the distance of y from its starting point y_0 cannot exceed t; that is, $\|y - y_0\| \leqslant t$. Hence, by condition (b), which says $\|y - y_0\| \leqslant t \Rightarrow \|S'(y)\| \leqslant \phi'(t)$, we know that $\|S'(y)\| \leqslant \phi'(t)$ during this motion. With the help of an inequality given at the end of section 6.5 we now have

$$\|S(y_1) - S(y_0)\| = \left\| \int_0^{t_1} S'(y)y'(t)\, dt \right\| \leqslant \int_0^{t_1} \|S'(y)\|\,\|y'(t)\|\, dt \leqslant \int_0^{t_1} \phi'(t)\, dt$$

since, as we saw earlier, $\|y'(t)\| \leqslant 1$. Finally

$$\int_0^{t_1} \phi'(t)\, dt = \phi(t_1) - \phi(t_0) = t_2 - t_1.$$

Thus, starting from $\|y_2 - y_1\|$, we have reached $t_2 - t_1$ by a sequence of steps involving $=$ and $\leqslant$. This proves what we require for the second links.

For the third links we proceed similarly. As $\|y_3 - y_2\| = \|S(y_2) - S(y_1)\|$, we arrange for the point y to continue its travels, moving with constant velocity from y_1 at t_1 to y_2 at t_2. We have proved that the length of the second link is less than $t_2 - t_1$, so here again the speed cannot exceed 1. Accordingly, between times $t = 0$ and $t = t_2$, the point y has never moved with speed more than 1, accordingly its distance from its starting point at no time exceeds t. Thus the stipulation in condition (b), 'so long as $\|y - y_0\| \leqslant t$' is satisfied, and we are entitled to conclude that $\|S'(y)\| \leqslant \phi'(t)$ still applies. By the same kind of argument as before we then show $\|y_3 - y_2\| \leqslant t_3 - t_2$.

It is clear that this construction can be continued indefinitely, the speed of y never exceeding unity, and the result for each interval providing the data to establish this for the next. Thus the correctness of the test can be proved by mathematical induction. The information about the chain $y_0, y_1, y_2,, \ldots$ so obtained permits us to conclude that the iteration does converge to some limit Y, to deduce that $\|Y - y_0\| \leqslant T$, and to estimate the rate at which this iteration converges.

7. Clearly the test can only be applied after a suitable function ϕ has been constructed, and we return to our particular example to illustrate how this is done.

From equations (11) and (12),

$$y_1(x) - y_0(x) = x/(x+1)^2.$$

The quantity $x/(x+1)^2$ increases until $x = 1$ and thereafter decreases. If we confine ourselves to the interval $[0, a]$ for x, with $0 < a \leqslant 1$, then $\|y_1 - y_0\| = a/(a+1)^2$ in the norm for $\mathscr{C}[0, a]$. Condition (a) requires $\|y_1 - y_0\| \leqslant \phi(0)$ and we satisfy this most economically by taking $\phi(0) = a/(a+1)^2$.

Condition (b) involved $S'(y)$, which we must now compute. The operator S is defined by equation (10). We ignore the subscript n in that equation, since we are concerned with $S(y)$, not $S(y_n)$. By considering $S(y+h) - S(y)$ we find

$$S'(y)h \cdot (x) = (x+1)^{-2} \int_0^x [2(s+1) - 2(s+1)^2 y(s)]h(s)\, ds.$$

We want $\|S'(y)\|$. By reasoning similar to that in item 5 of section 5.4 we find

$$\|S'(y)\| = \sup\left\{(x+1)^{-2} \int_0^x |2(s+1) - 2(s+1)^2 y(s)|\, ds\right\}. \qquad (14)$$

The supremum here indicates that we are to take the value of x in $[0, a]$ that makes the quantity on the right-hand side a maximum.

As we do not know what function y represents, this appears a rather intractable expression and it is not clear what we should do next. The clue is provided by the wording of condition (b); we are seeking a function ϕ such that $\|y - y_0\| \leqslant t \Rightarrow \|S'(y)\| \leqslant \phi(t)$.

Now $y_0(s) = 1/(s+1)$, so $\|y - y_0\| \leqslant t$ means

$$|(s+1)^{-1} - y(s)| \leqslant t.$$

Multiplying by $(s+1)^2$ shows

$$|2(s+1) - 2(s+1)^2 y(s)| \leqslant 2t(s+1)^2.$$

In this inequality we observe the integrand of equation (14). Accordingly, when $\|y - y_0\| \leqslant t$ we have

$$\|S'(y)\| \leqslant \sup\left\{(x+1)^{-2}\int_0^x 2t(s+1)^2 \, ds\right\}.$$

As

$$\int_0^x (s+1)^2 \, ds = [(x+1)^3 - 1]/3 < (x+1)^3/3,$$

we find $\|S'(y)\| \leqslant \sup\{2t(x+1)/3\} = 2t(a+1)/3$ since we suppose x to be in $[0, a]$.

If we take $\phi'(t) = 2t(a+1)/3$ we ensure that condition (b) is satisfied.

Thus

$$\phi(t) = \frac{a}{(a+1)^2} + \frac{a+1}{3} t^2 \tag{15}$$

meets both conditions.

The graph of ϕ is a parabola. If we take a very small, the parabola will be almost flat and the iteration $t_n \to \phi(t_n)$ will converge rapidly. We know the iteration $y_n \to S(y_n)$ will do at least as well. However our information will be for x in a very short interval, $[0, a]$, only. With larger values of a, we shall know what happens for a longer interval, but the rate of convergence we shall be able to prove must be less. It should be remembered that our analysis so far has been based on the assumption that $a \leqslant 1$. For $a > 1$ we would have to consider $\phi(t) = 0.25 + [t^2(a+1)/3]$. There is a natural barrier to our investigation at $a = 2$. If $a > 2$, the equation $t = \phi(t)$ has no real root and the iteration $t_n \to \phi(t_n)$ cannot converge. This does not prove that the iteration $y_n \to S(y_n)$ diverges for $x > 2$. It merely means that our present method of proving its convergence evaporates.

We have carried the iteration only as far as y_2. What information does Kantorovich's test give us about $\|y_2 - Y\|$, the distance of y_2 from the exact solution Y? Since the chain $y_0, y_1, y_2, \ldots$ approaches the limit Y, the distance of y_2 from Y cannot exceed the (infinite) sum of the lengths of all the links that come after y_2. By our theorem, this cannot exceed the sum of the lengths of all the links that come after t_2 in the sequence $0, t_1, t_2, \ldots$. These links fill the interval between t_2 and T; the sum of their lengths is $T - t_2$. Accordingly $\|y_2 - Y\| \leqslant T - t_2$.

The table below gives the values of $T - t_2$ for various values of a, with $t_{n+1} = \phi(t_n)$ and $\phi(t)$ as given by equation (15). Now Y, the solution of equation (5), can be obtained analytically by writing $y(x) = z(x)/(x+1)$ and separating the variables x and z. The values of $Y - y_2$ are calculated and given for comparison in the following table. All entries are to two significant figures.

a	$T - t_2$	$\lVert Y - y_2 \rVert$
0.1	0.00016	0.000 000 83
0.2	0.0010	0.000 017
0.3	0.0026	0.000 086
0.4	0.0049	0.000 25
0.6	0.011	0.000 93
0.8	0.018	0.002 1
1.0	0.025	0.003 6

It will be seen that the actual iteration does much better than is indicated by the value of $T - t_2$. This is to be expected. Any general theorem has to cover the worst possible situation consistent with the assumptions. This can be seen in our work above. The distance $y_2 - Y$ will equal the sum of the lengths of the links between y_2 and Y only if all of these are in roughly the same direction, which is unlikely. (We cannot say 'in exactly the same direction' as we would for Euclidean space, because the space of continuous functions is like the chessboard with the king's move.) Again, when we use $\int_0^1 |f(x)|\, dx$ as the norm of the function

$$y \rightarrow \int_0^1 f(x) y(x)\, dx,$$

we are taking account of the fact that there could be a function y, with $\lVert y \rVert = 1$, diabolically adapted to bring out all the worst in the function f. Such a function y would have the value $+1$ for $f(x) > 0$, but whip over as quickly as possible to -1, so soon as $f(x)$ became negative. It would be bad luck indeed if, in any concrete example involving $\int_0^1 f(x) y(x)\, dx$, the particular function y behaved like this.

As the aim has been to sketch ideas rather than to give complete formal proofs, it may be useful now to state the conditions required by the above test. The operator S must have a continuous derivative, $S'(y)$, for all y in the closed ball $\bar{B}(y_0, T)$; ϕ must be defined and differentiable in $[0, T]$; the conditions (a) and (b) must be satisfied.

So far we have not required the iteration $y_n \to S(y_n)$ to be an iteration arising from the Newton–Raphson procedure. Kantorovich developed this idea and created a theory for the situation where S is such an iteration, defined as in equation (9) of this chapter. The theory involves the second derivative, $f''(y)$, which we have not yet considered. An account of this theory will be found in chapter XVIII of Kantorovich and Akilov.

Exercises

In these questions the symbols, f, S and ϕ, have the meanings attached to them in section 6.6.

1. Find the operator S corresponding to the function $f: \mathbb{R}^2 \to \mathbb{R}^2$, $(x, y) \to (xy + 0.07, x^2 + y^2 - 0.41)$, with the initial $(x_0, y_0) = (0.1, -0.6)$.

Show that $S'(0.1 + X, -0.6 + Y)$ is given by a 2×2 matrix, M, with $m_{11} = m_{22} = (2X + 12Y)/7$ and $m_{12} = m_{21} = (12X + 2Y)/7$. Hence show that $|X| \leq t, |Y| \leq t$ implies $\|S'(0.1 + X, -0.6 + Y)\| \leq 4t$. Deduce that the iteration, $t \to \phi(t) = 0.0315 + 2t^2$, may be used for comparison with the modified Newton–Raphson iteration. Verify this by carrying out both iterations.

Compare this method with that used at the beginning of section 6.4.

2. For $f: \mathbb{R} \to \mathbb{R}$, $(x, y) \to (x^2 + y^2 - 200, y^3 + xy - x^3)$ and the initial $(x_0, y_0) = (10, 10)$ show that $[f'(x_0, y_0)]^{-1} = N$, where

$$N = (1/12000)\begin{pmatrix} 310 & -20 \\ 290 & 20 \end{pmatrix}.$$

Show that

$$S'(x, y) = I - N\begin{pmatrix} 2x & 2y \\ y - 3x^2 & 3y^2 + x \end{pmatrix}.$$

What is $S'(x_0, y_0)$? Is this result an accident or will it always happen with the Newton–Raphson method? Find $S'(10 + X, 10 + Y)$ and show that $\|S'(10 + X, 10 + Y)\|_\infty$ does not exceed $(t/3) + (t^2/100)$ when $|X| \leq t$ and $|Y| \leq t$. (This can be shown by a crude approximation in which every term is replaced by its absolute value.) Hence show that the comparison function ϕ can be defined by $\phi(t) = (1/6) + (t^2/6) + (t^3/300)$. Check this by carrying out the iterations both of S and of ϕ.

3. The function $f: \mathbb{R}^n \to \mathbb{R}^n$ is defined by $f(v) = Mv + g(v)$ with constant matrix M. The equation, $f(v) = 0$ is to be solved by the Newton–Raphson method with the initial vector, v_0. It is given that $g'(v_0) = 0$. Prove that S is the function $v \to -M^{-1}g(v)$.

Find S for the function $f:\mathbb{R}^2\to\mathbb{R}^2$, $(x, y)\to(-37x+9y+x^5+y^5+25, 4x-28y+x^3y^3+18)$, with $(x_0, y_0) = (0, 0)$. Show that, when $|x|\le t$ and $|y|\le t$, then

$$\|S'(x, y)\|_\infty \le 0.28t^4 + 0.222t^5.$$

Deduce that a possible comparison function, ϕ, is given by

$$\phi(t) = 0.862 + 0.056t^5 + 0.037t^6.$$

Carry out the iterations both of S and ϕ. (Note. For the relevant values of t, an improved choice of ϕ can be found.)

4. The function $S:\mathscr{C}[0, a]\to\mathscr{C}[0, a]$, where $a > 0$, is defined by $y\to z$ with

$$z(x) = \int_0^x [y(t)+t]^2\, dt.$$

Show that, with $y_0(x)\equiv 0$, the behaviour of the iteration of S can be estimated by means of the comparison function ϕ, with $\phi(t) = (a^3/3)+a^2t+at^2$.

Show that the iteration of S arises naturally from the Newton–Raphson procedure for solving the differential equation, $du/dx - (x+u)^2 = 0$, with the condition, $u(0) = 0$ and the initial $u_0(x) = -x$. Verify that $u_1 = y_0$.

5. The function $f:\mathscr{C}[0, 1]\to\mathscr{C}[0, 1]$, $y\to z$ has

$$z(x) = b - y(x) + ax[y(x)]^2 + a\int_1^x [y(s)]^2\, ds \quad \text{with} \quad a > 0, b > 0.$$

With initial $y_0(x)\equiv 0$, find the function S used to solve the equation $f(y) = 0$. (The inverse of $f'(y_0)$, often hard to find, is here immediately evident.)

Show that

$$S'(y_0)h\cdot(x) = 2axy(x)h(x) + a\int_1^x 2y(s)h(s)\, ds.$$

Deduce that $\phi(t) = b + at^2$ can be used in the Kantorovich comparison iteration.

For the case $a = 1$, $b = 0.09$, find y_1, y_2, y_3, and compare $\|y_{n+1} - y_n\|$ with $t_{n+1} - t_n$ for $n = 0, 1, 2$. What unusual thing happens in this example?

6. Show that, if $f(y) = Ly - g(y)$, where L is a linear operator and $g'(y_0) = 0$, then $S(y) = L^{-1}g(y)$.

Show that, if

$$h(x) + \int_0^x h(s)\, ds = k(x),$$

then

$$h(x) = k(x) - e^{-x} \int_0^x e^s k(s) \, ds.$$

Let $f: \mathscr{C}[0, c] \to \mathscr{C}[0, c], y \to z$ have

$$z(x) = y(x) + \int_0^x y(s) \, ds - a[y(x)]^2 - b.$$

The constants a, b, c are all positive. With the help of the results above, find S, the function iterated to solve the equation $f(y) = 0$, the initial y_0 having $y_0(x) \equiv 0$. Show that

$$S'(y)h \cdot (x) = 2ay(x)h(x) - 2a \, e^{-x} \int_0^x e^s y(s)h(s) \, ds.$$

Find a suitable function ϕ for the Kantorovich comparison iteration.

Compute y_1, y_2, y_3, for $a = 1$, $b = 0.09$. Do the results suggest any connection between this question and question 5 above?

7 Further developments

7.1. Taylor series

THE PURPOSE of this chapter is to indicate the existence of further analogies to results and methods, without going fully into their details. The Taylor series is the first of such items.

The Taylor series for a function $f: \mathbb{R} \to \mathbb{R}$, takes the form

$$f(x+h) = f(x) + f'(x)h + (1/2!)f''(x)h^2 + \ldots (1/n!)f^{(n)}(x)h^n + \ldots.$$

An evident problem arises in generalizing this; if h is to be a vector, how are we to cope with $h^2, h^3, \ldots$, none of which is defined in a vector space? The answer to this hinges on the nature of the successive derivatives. It will be remembered that, when f was a function, vector $\to$ vector, $f'(v)$ turned out to be not a vector but a matrix. A matrix resembles a computer program that demands a vector as input and produces a vector as output. When we come to higher derivatives for the same type of function, we find in effect that $f''(v)$ requires two vectors to be fed in and then gives a vector as output. In the same way, $f^{(n)}(v)$ feeds on n vectors before it yields its output of a single vector. Thus, instead of $f''(x)h^2$, in which h is multiplied by itself, we have to think in terms of $f''(v) \cdot h \cdot h$, where the $h \cdot h$ indicates that the vectors fed into the two input slots of $f''(v)$ are both to be given the value h. A similar understanding is to operate when we generalize the other terms of the Taylor series.

We may illustrate how this works by considering a function $f: \mathbb{R}^2 \to \mathbb{R}$. The output of f is now a single real number, which can be regarded as a vector in one dimension. In $\mathbb{R}^2$, we may take $h = (a, b)$, so this simple example is quite sufficient to illustrate the non-existence of h^2, and the way in which this obstacle is overcome. Let $f(x, y) = 3x^2 + 4xy + 5y^2$. Then to find f' we follow the procedure described in section 6.2, and write

$$f(x+a, y+b) - f(x, y) = (6x+4y)a + (4x+10y)b + \ldots.$$

Since the right-hand side here may be written as

$$(6x+4y, 4x+10y)\begin{pmatrix} a \\ b \end{pmatrix}$$

with the usual row-and-column rule for multiplication, we may think of $f'(x, y)$ as being specified by the row vector $(6x + 4y, 4x + 10y)$. If now (x, y) again changes to $(x + a, y + b)$, the derivative $f'(x, y)$ changes by $(6a + 4b, 4a + 10b)$, or $(6, 4)a + (4, 10)b$. If we keep the convention of row-and-column multiplication, but permit vectors to appear as entries, we can write

$$f''(x, y)\binom{a}{b} = [(6, 4), (4, 10)]\binom{a}{b}$$

so $f''(x, y) = [(6, 4), (4, 10)]$, a type of expression that is not usual in elementary matrix and vector work. Two vectors have to be fed into $f''(x, y)$ before a real number results. A meaningful expression would be reached with two different vectors, as in

$$f''(x, y)\binom{a}{b}\binom{c}{d}.$$

To evaluate this, we would suppose the vectors acted on by $f''(x, y)$, one after the other.

$$f''(x, y)\binom{a}{b} = [(6, 4), (4, 10)]\binom{a}{b} = (6, 4)a + (4, 10)b$$
$$= (6a + 4b, 4a + 10b).$$

Then

$$(6a + 4b, 4a + 10b)\binom{c}{d} = 6ac + 4bc + 4ad + 10bd.$$

For Taylor's series, it is not necessary to consider distinct vectors; we have $(c, d) = (a, b)$ so the term $(1/2!)f''(x, y)(a, b)(a, b)$ is $(1/2) \times (6a^2 + 8ab + 10b^2) = 3a^2 + 4ab + 5b^2$. The previous term, $f'(x, y)(a, b)$ is $(6x + 4y)a + (4x + 10y)b$. As f'' is constant, all higher derivatives $f^{(n)}(x, y)$ are 0, so the Taylor series reduces to

$$f(x, y) + f'(x, y)(a, b) + (1/2)f''(x, y)(a, b)(a, b)$$
$$= 3x^2 + 4xy + 5y^2 + (6x + 4y)a + (4x + 10y)b + 3a^2 + 4ab + 5b^2.$$

In this simple case, of course, the Taylor series tells us nothing that we could not have found, more easily, by elementary algebra. We have considered it only to show the nature of the successive derivatives, $f'(x, y)$, a vector with two entries, and $f''(x, y)$, an entity with four entries. If the problem had involved the third derivative, $f'''(x, y)$, that would have had eight entries, and so on.

It might be thought that the proof of Taylor's theorem in a very general situation would be long and difficult. In fact, the proof is very short and simple. Taylor's theorem, with a remainder term, can be proved by exactly the same method as was used in section 6.3 to prove the mean value theorem. A linear operator, L, with the four properties required there, is brought in, and the theorem quickly reduces to one in traditional analysis, $\mathbb{R} \to \mathbb{R}$. (See Collatz, section 16.6.)

Many students find it takes time to get used to the way in which the complexity of the nth derivative, $f^{(n)}(v)$, increases as n increases. It may therefore be worth mentioning that Kantorovich's treatment of the Newton–Raphson method does not involve any derivative of order higher than the second. It is thus possible to read chapter XVIII of Kantorovich and Akilov once the meaning of the first two derivatives has been grasped. This provides a breathing space, during which the concepts of the higher derivatives can gradually sink into the mind and become familiar.

7.2. Differentiation and linear functions

If T is a bounded linear operator, the equation $w = Tv$ has many analogies with the equation for real numbers, $y = mx$. If v changes to $v + h$, then $\Delta w = T(v + h) - Tv = (Tv + Th) - Tv = Th$, so $T'(v) = T$, just as for $f(x) = mx$, $f'(x) = m$.

Again, for any differentiable function ϕ, $(d/dx)m\phi(x) = m\phi'(x)$. The m slips, unaltered, past the differentiation operator. Similarly, if

$$u = T\Phi(v), \qquad \Delta u = T\Phi(v + h) - T\Phi(v)$$
$$= T[\Phi(v + h) - \Phi(v)] = T[\Phi'(v)h + e(h)]$$
$$= T\Phi'(v)h + Te(h),$$

where, as usual, $e(h) \to 0$ with h. As T is bounded, $\|Te(h)\| \leqslant \|T\| \cdot \|e(h)\|$, so $Te(h)$ also tends to 0, and so $(T\Phi)' = T \cdot \Phi'$. This result was used in section 6.3.

It is a curious fact, that difficulties of notation are most likely to arise in connection with the simplest ideas. Differentiation of linear operators is a case in point. There is a great temptation, in view of the equation $T'(v) = T$ to write simply $T' = T$. But that would lead us to write $T'' = (T')' = T' = T$, and it would seem that all derivatives of T were identical. But this is something we associate with $f(x) = e^x$ rather than $f(x) = mx$. We know it cannot be true, for $x \to mx$ is itself an example of a bounded, linear function, and for it $f'' = 0$, which is not the same as f' or f.

The confusion arises from the fact that two variables are involved in differentiation. If $F(x) = x^2$, $\Delta F(x) = 2xh + \dots$ and $F'(x) = 2x$. The second derivative, F'', is not 0 because the first derivative, $F'(x)$, grows when x does. Now for $f(x) = mx$, $\Delta f(x) = mh$ and $f'(x)$ is the function $h \to mh$. Certainly this is of the same form as $x \to mx$. The point however is that this function does not involve x; the line $y = mx$ has the same slope at every point, so $f''(x) = 0$. To go from $f'(x)$ to $f''(x)$, we apply the operation d/dx, not d/dh. It is the same for the function $v \to Tv$. The derivative is $h \to Th$, of the same form, but $T'(v) = T$ and T is constant, independent of v. Accordingly $T''(v) = 0$. It is clear from the Taylor series that this must be so.

As a simple numerical example, consider the matrix

$$T = \begin{pmatrix} 2 & 3 \\ 4 & 5 \end{pmatrix}$$

so that we have

$$T : (x, y) \to (2x + 3y, 4x + 5y).$$

If (x, y) changes to $(x + a, y + b)$, the output changes by $(2a + 3b, 4a + 5b)$. Accordingly

$$T' \begin{pmatrix} x \\ y \end{pmatrix} = \begin{pmatrix} 2 & 3 \\ 4 & 5 \end{pmatrix}.$$

The matrix for $T'(x, y)$ does not involve x and y, so $T''(x, y) = 0$.

Note that $T'(v)$ does not mean $T'v$, the operator T' acting on v. Rather $T'(v)$ means the value of the derivative taken at the point v, exactly as in $f'(x)$. We can then consider $T'(v)h$, this operator acting on h, where h is the change in v.

7.3. Majorants

In section 6.3 it was shown that the iteration with

$$f_{n+1}(x) = x + \int_0^x [f_n(t)]^2 \, dt$$

gave a contraction mapping for $0 \leq x \leq a$, provided $2af(a) < 1$. It was mentioned that a suitable value of a could be found without actually carrying out the iteration.

It is clear that the expression, on the right-hand side of the equation above, increases if x increases. For the integral of any positive function from 0 to x must grow if x grows, and the first term, x, also grows. So the functions we obtain as the iteration proceeds will all be monotonic increasing. Suppose b is a number such that $f_n(a) \leq b$. Then, in the interval of integration, $f_n(t) \leq b$, so $f_{n+1}(a) \leq a + ab^2$. Choose b as a solution of the equation $b = a + ab^2$. If $f_n(a) \leq b$, it will follow that $f_{n+1}(a) \leq b$. Taking $f_0(0) \equiv 0$ will ensure $f_0(a) \leq b$, provided b is a positive number and allow us to prove $f_n(a) \leq b$ for all n by induction. We will suppose $0 < a \leq 0.5$, and choose for b the smaller root of the equation. This is $[1 - \sqrt{(1 - 4a^2)}]/(2a)$, and gives a real positive value for b.

Our earlier work showed that the contraction factor for the step $f_n \rightarrow f_{n+1}$ did not exceed $2af_n(a)$, which we now see does not exceed $2ab$. Accordingly, from the value of b above, the contraction factor is not more than $1 - \sqrt{(1 - 4a^2)}$. Taking $a = 0.5$ would give the value 1, which is not helpful. Taking $a = 0.4$ makes the estimate for the contraction factor also 0.4, with quite rapid convergence.

As $f_n(x) \rightarrow f(x)$, our inequalities $f_n(a) \leq b$ imply $f(a) \leq b$ and so give estimates for the actual solution. This solution in fact is $f(x) = \tan x$. The following table gives b for various values of a, and the value of $\tan a$ for comparison. The last column gives $2ab$, our estimate for the contraction factor in the iteration.

a	b	$\tan a$	$2ab$
0.1	0.1010	0.1003	0.0202
0.2	0.2087	0.2027	0.0835
0.3	0.3333	0.3093	0.2
0.4	0.5	0.4228	0.4
0.5	1.0	0.5463	1.0

Figure 34 shows the graph of $z(t) = a(1 + t^2)$. The number b corresponds to the first intersection with the line $z = t$. The gradient of the graph at this intersection is $2ab$, which gave our estimate for the contraction factor. With $a = 0.5$ the parabola would just touch the line.

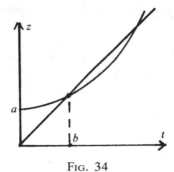

Fig. 34

7.4. Polynomial operators

The convergence for $x < 0.5$ of the iteration considered in section 7.3 could have been established by the method of Kantorovich, with $\phi(t) = a(1 + t^2)$, the same function as that graphed in Figure 34. The result in section 7.3 can also serve as an introduction to the work of Rall on polynomial operators. That result can in fact be seen as a particular case of a general theorem.

First of all it is necessary to consider what is meant by a polynomial operator and its degree. Consider this first in spaces of finite dimension. If $w_r = \sum_s a_{rs} u_s$, we regard the function $A : u \to w$ as linear. For the function $(u, v) \to w$ defined by

$$w_r = \sum_s \sum_t c_{rst} u_s v_t,$$

the mapping is linear in regard to u if v is held fixed, and linear in regard to v if u is fixed. Such a mapping is called bilinear. In section 7.1, when we considered $f''(x, y)(a, b)(c, d)$, the mapping $f''(x, y)$ was bilinear.

The mapping $(u, v) \to w$ can be broken into two stages. If we write $m_{rt} = \sum_s c_{rst} u_s$, then $w_r = \sum_t m_{rt} v_t$. We may write $w = Mv$, where M is the matrix with entries m_{rt}. In the first stage, specified by $m_{rt} = \sum_s c_{rst} u_s$, the input is the vector u, the output is the matrix M. We may write $M = Cu$, where C is a function, vector $\to$ matrix. Thus $w = Mv = (Cu)v$. Usually this is written $w = Cuv$, the bracketing being understood. The expression must be read from left to right. The operator C can only get at u, which is written next to it. Acting on u, C produces $Cu = M$, an operator capable of acting on v. The final output is $Mv = w$.

Both stages of the above process involve bounded, linear functions. The operator norm is defined for both stages. As $M = Cu$, $\|M\| \leqslant \|C\| \|u\|$. As $w = Mv$, $\|w\| \leqslant \|M\| \|v\|$. Combining these, we have $\|w\| \leqslant \|C\| \|u\| \|v\|$. This last inequality gives us another way of looking at the norm, $\|C\|$. We can define $\|C\|$ as the infimum of the numbers k for which $\|w\| \leqslant k \|u\| \|v\|$, where $w = Cuv$. This formulation is often convenient.

When we leave finite dimensional spaces and consider general Banach spaces, we can no longer write explicit algebraic formulas for the mappings involved, but the general run of ideas is the same. If the vectors u, v, w were in the spaces $\mathcal{U}, \mathcal{V}, \mathcal{W}$ respectively, the matrix M served simply to specify an operator in $\mathcal{B}(\mathcal{V}, \mathcal{W})$. The equation $Cu = M$ shows that C acted on a vector u in U, and the output was the matrix M in $B(V, W)$. To generalize this, we say that C is a function $\mathcal{U} \to \mathcal{B}(\mathcal{V}, \mathcal{W})$.

In one important class of applications, we are still very close to our simple matrix notation. The integral equation

$$w(x) = \int_0^1 \int_0^1 \phi(x, y, z) u(y) v(z) \, dy \, dz,$$

involving a continuous function ϕ, resembles the equation

$$w_r = \sum_s \sum_t c_{rst} u_s v_t,$$

with integration replacing summation. We can break this mapping into two stages by writing

$$m(x, z) = \int_0^1 \phi(x, y, z) u(y) \, dy$$

and

$$w(x) = \int_0^1 m(x, z) v(z) \, dz.$$

In the first stage, the input function u leads to the expression $m(x, z)$, needed to define the bounded, linear operator $M : v \to w$, that is used in the second stage.

It is possible to continue and explain what is meant by $Pzuv$, an expression involving three vectors, z, u and v. If we write $Pz = C$, then $Pzuv = Cuv$. We have already seen what kind of operator C makes Cuv meaningful, so P must accept a single vector u as an input and deliver an operator of that kind as its output. We could continue indefinitely in this way, as was briefly indicated when Taylor series were considered.

Polynomials are obtained by making all the input vectors identical. If we put $u = v = h$ in Cuv, we obtain Chh, which counts as an expression of the second degree. Similarly from $Pzuv$ we can form $Phhh$, of the third degree. By the most general cubic polynomial in h we would understand an expression of the form $a + Bh + Chh + Phhh$, where P and C are operators as described above, B is a bounded linear operator, and a is a fixed vector. The vector a is constant in the sense that it does not depend on h. If the vector a represents a function, this does not mean that that function must be a constant. For example, if $a \in \mathscr{C}[0, 1]$, a could be the function $x \to x^2$ or any other continuous function defined without reference to h.

The examples in section 7.5 involve only quadratic polynomials.

7.5. Quadratic majorants

We consider the equation $v = q + Cvv$. Here q is a fixed vector in a Banach space, and we wish to find the unknown vector, v, in the same space. The operator C is given, and of course must be such as to make the equation meaningful.

A quadratic function, $\mathbb{R} \to \mathbb{R}$, $t \to p + ct^2$, is said to majorize the quadratic function, $v \to q + Cvv$ if $p \geq \|q\|$ and $c \geq \|C\|$. When this is so, we can make a useful comparison between a series solution for v in the Banach space and a series solution for the real number t.

Consider the equation $t = p + kct^2$ and assume that, for sufficiently small, positive k, it can be satisfied by a series $t = \sum_0^\infty t_r k^r$. Then

$$t_0 + t_1 k + t_2 k^2 + t_3 k^3 + \ldots = p + kc(t_0 + t_1 k + t_2 k^2 + t_3 k^3 + \ldots)^2.$$

Equating coefficients of powers of k, we have

$$t_0 = p, \quad t_1 = ct_0^2, \quad t_2 = c(2t_0 t_1), \quad t_3 = c(2t_0 t_2 + t_1^2),$$

and generally

$$t_n = c \sum_{r=0}^n t_r t_{n-r}.$$

As the multiplication of infinite series can lead to paradoxical results, the calculations just made have to be justified. The squaring of a convergent series, $\sum_0^\infty a_r$, by the same algorithm as for finite sums is justified if either (i) for all r, $a_r \geqslant 0$, or (ii) the series is absolutely convergent, that is, $\sum |a_r|$ is convergent. (See Knopp, *Theory of Infinite Series*, sections 16 and 17.)

Condition (i) clearly holds for our work above, since each term $t_r k^r$ is clearly not negative. A generalization of condition (ii) is relevant to the calculation that will be made in a moment. The operations to be performed are justified if the series, $\sum a_r$, is absolutely convergent in the sense that $\sum \|a_r\|$ is convergent.

We proceed similarly with our vector equation, $v = q + kCvv$, by assuming that it has a series solution $v = \sum_0^\infty v_r k^r$. We treat it in the same way, with care about the order in which symbols are written, for Cuv may not mean the same thing as Cvu, and uCv or uvC is meaningless. The result is

$$v_0 = q, \quad v_1 = Cv_0 v_0, \quad v_2 = Cv_0 v_1 + Cv_1 v_0,$$

$$v_3 = Cv_0 v_2 + Cv_1 v_1 + Cv_2 v_0,$$

and so on. With the help of the majorizing conditions, we work through these in turn. First

$$\|v_0\| = \|q\| \leqslant p = t_0.$$

Then

$$\|v_1\| = \|Cv_0 v_0\| \leqslant \|C\| \|v_0\| \|v_0\| \leqslant cp^2 = t_1.$$

It is possible to continue in this way and to show $\|v_r\| \leqslant t_r$ for each r. Accordingly, if $\sum t_r k^r$ converges, the series, $\sum v_r k^r$, converges absolutely and the formal steps taken are justified.

So the equation $v = q + kCvv$ has a solution, given by a convergent series, provided the same is true for the majorant equation $t = p + kct^2$. The series for t found above corresponds to the solution $t = [1 - \sqrt{(1 - 4kpc)}]/(2kc)$ of the majorant equation. The function of k involved here has a singularity at $k = 1/(4pc)$. It is known from the theory of functions of a complex variable that the series for an analytic function converges within the circle, that has its centre at the origin and passes through the nearest singularity. Accordingly the series converges for positive k such that $4kpc<1$. It will converge for $k=1$ if $4pc<1$. Putting $k=1$ brings us back to our original equation, $v=q+Cvv$. It follows that this equation will have a convergent series solution, provided $\|q\|\,\|C\|<0.25$.

An example of such an equation has already been met. At the beginning of section 7.3 we had the equation

$$f(x)=x+\int_0^x [f(s)]^2\, ds,$$

in an equivalent form. This involves a quadratic polynomial in f, for if

$$Cuv\,(x)=\int_0^x u(s)v(s)\, ds,$$

then C is bilinear, and putting $u=v=f$ gives the integral occurring in the equation. The term x corresponds to the 'constant' q. If we are considering x in $[0, a]$, $\|q\|=a$ and $\|C\|=a$. Thus the majorant quadratic is $a+at^2$. This is the same quadratic that was used for comparison in section 7.3, though the assertions made now are not exactly the same as those made then.

A celebrated application of this theory is to Chandrasekhar's equation for radiative transfer,

$$f(x)=1+kf(x)\int_0^1 \frac{xf(y)}{x+y}\, dy.$$

This application is quite remarkable. The equation looks complicated, and it surely would not occur to many people on first seeing this equation to classify it with anything as simple as the traditional quadratic equation. In the literature it is often met with less familiar symbols, μ and μ' instead of x and y, and $\omega_0/2$ instead of k. This alteration, of course, is purely superficial and trivial, but it can have the psychological effect of making the equation look even more mysterious.

First of all, the equation is a quadratic. If

$$Cuv\cdot(x)=ku(x)\int_0^1 \frac{xv(y)}{x+y}\, dy,$$

then C is a bilinear operation. We suppose $f\in\mathscr{C}[0, 1]$. To find $\|C\|$, we suppose $\|u\|=\|v\|=1$ prescribed. Then the largest values for

$Cuv \cdot (x)$ are obtained by taking $u(x) \equiv 1$, $v(x) \equiv 1$. Then

$$Cuv(x) = k \int_0^1 \frac{x}{x+y} \, dy = kx[\ln(x+1) - \ln x].$$

This has its maximum value, $k \ln 2$, when $x = 1$. Thus $\|C\| = k \ln 2$. As $\|q\| = 1$, the convergence condition becomes $k \ln 2 < 0.25$. Thus there is a solution of the integral equation, given by a convergent series, provided $k < 0.36067$.

An account of this approach, with some more general theory in which Taylor series are considered instead of just quadratics, will be found in chapter 15 of Delves and Walsh. This chapter is written by L. B. Rall. More detailed theory can be found in Rall (1, 2) and Prenter.

It may be noticed that a common thread links three approaches that have been described—contraction mappings, the Kantorovich test, Rall's quadratics. In each case it is shown that some process in a general space works at least as well as an associated process with real numbers.

8 Euclidean space

8.1. On perpendicularity

IN THE SPACES we have considered so far, the concept 'perpendicular' has not been used. This is not surprising; we associate perpendiculars with the shortest path from a point to a line. But in many spaces, no point is singled out by the property of nearness. The first hint of this was in the chessboard diagram of chapter 2; all the points on the right-hand side of the board were at the same distance, 4, from the king, yet these points were in line. The ℓ_∞ space has some kinship with the king's move metric. In it, the unit sphere is a square, and all the points on one side of the square are in line and are the same distance from the centre. The important space $\mathscr{C}[a, b]$ shows a similar effect.

Exercise. The functions $u + mv$, where u and v are fixed functions and the number m varies, form a straight line in $\mathscr{C}[0, 1]$. If $u(x) \equiv 1$, $v(x) \equiv x$, find for what values of m the distance, in the metric of $\mathscr{C}[0, 1]$, between $u + mv$ and the function 0, takes its minimum value.

There are certain situations in functional analysis in which 'perpendicular' can be defined. In these, great simplifications occur and surprising applications can be made. It will be useful, as a preparation for later generalization to review the theory of perpendiculars in Euclidean space of three dimensions. The work will be in terms of co-ordinates, based on the three-dimensional system that corresponds to squared paper in two dimensions; that is, we imagine space divided up by a great assemblage of little cubes.

If the vectors $u = (u_1, u_2, u_3)$ and $v = (v_1, v_2, v_3)$ give the co-ordinates of the points U and V, the distance $\|UV\|$ is given by

$$\|UV\|^2 = (v_1 - u_1)^2 + (v_2 - u_2)^2 + (v_3 - u_3)^2 = \sum (v_r - u_r)^2.$$

If $OU \perp OV$, Pythagoras' Theorem indicates that $\|OU\|^2 + \|OV\|^2 = \|UV\|^2$. Algebraically this means $\sum u_r^2 + \sum v_r^2 = \sum (v_r - u_r)^2$, which reduces to $\sum u_r v_r = 0$. The quantity $\sum u_r v_r$ is known as the scalar (or 'dot') product of u and v. In elementary work it is usually denoted by $u \cdot v$, and in more advanced work by (u, v). If complex numbers

are accepted as co-ordinates the definition has to be slightly modified; we shall not be concerned with this. The term 'product' is used
because this expression has many of the formal properties of a
product. For instance, if a and b are numbers, $u \cdot (av + bw) =$
$a(u \cdot v) + b(u \cdot w)$.

In the present section, the symbol $u \cdot v$ will be used, but in later
sections, when generalizations are considered, the form (u, v), that is
usual in the literature, will be employed.

The scalar product arises naturally in other circumstances, besides
expressing $u \perp v$. For instance, if OU and OV are at an angle θ, we
may enquire what point of the line OV is nearest to U. Any point
on the line OV is of the form tv. The distance of this point from U
is d, where

$$d^2 = \sum (u_r - tv_r)^2 = \sum u_r^2 - 2t \sum u_r v_r + t^2 \sum v_r^2$$
$$= (u \cdot u) - 2t(u \cdot v) + t^2(v \cdot v).$$

This result could have been obtained without going back to the
formula for distance. As, for any vector w, $\|w\|^2 = \sum w_r^2 = w \cdot w$, we
can express the desired distance in terms of scalar products, and
then use their algebraic properties. Thus we have

$$d^2 = \|u - tv\|^2 = (u - tv) \cdot (u - tv)$$
$$= (u \cdot u) - 2t(u \cdot v) + t^2(v \cdot v).$$

Whichever way we arrive at this expression, we can find its
maximum by differentiating. The minimum occurs for $t_0 =$
$(u \cdot v)/(v \cdot v)$. Let M be the point for which this least distance
occurs. On geometrical grounds we expect $MU \perp OV$. The vector
MU is $u - t_0 v$. It will be perpendicular to OV if $0 = (u - t_0 v) \cdot v =$
$(u \cdot v) - t_0(v \cdot v)$, and from the value of t_0 already found we see that
this will be so. Thus we have an algebraic proof of the connection
between perpendicularity and nearness. Algebraic proofs are important because, while we may guess that certain things will happen in
Euclidean space of n dimensions, where $n > 3$, by analogy with what
happens in three dimensions, the physical universe does not give us
the same direct insight into n dimensions that it does into three.
Usually we have to rely on algebra to obtain proof.

As OM is the projection of OU onto OV, we naturally expect
OM to have a length not more than that of OU. That is, we expect

$\|t_0 v\| \leqslant \|u\|$, which means $|t_0| \|v\| \leqslant \|u\|$. Now $t_0 = (u \cdot v)/(v \cdot v) = (u \cdot v)/\|v\|^2$. Substituting and simplifying, we obtain $|(u \cdot v)| \leqslant \|u\| \cdot \|v\|$. We have reached this inequality on geometrical grounds. Its algebraic proof and its generalizations will be considered later.

The formula for the projection, OM, is particularly simple when v is a vector of unit length. Then $\|v\| = 1$, $t_0 = (u \cdot v)$ and $OM = (u \cdot v)v$.

We have considered the shortest distance from a point to a line. It is also possible to consider the shortest distance from a point to a plane through the origin. We can get useful information by considering the very simplest case, the distance of the point (u_1, u_2, u_3) from the plane $z = 0$. Any point of the plane is of the form $(x, y, 0)$ and the distance d from U is given by $d^2 = (u_1 - x)^2 + (u_2 - y)^2 + u_3^2$. It is evident algebraically, as it is geometrically, that we make d a minimum by taking $x = u_1$, $y = u_2$, so the nearest point to U is $(u_1, u_2, 0)$. This is the same as $(u_1, 0, 0) + (0, u_2, 0)$. The first of these is the point in line with the vector $(1, 0, 0)$ that is closest to U; the second is the point in line with $(0, 1, 0)$ that is closest to U. This can be seen geometrically. It is also simple to check algebraically, as $(1, 0, 0)$ and $(0, 1, 0)$ are both vectors of unit length, and the simple formula for the projection applies to them. Accordingly the nearest point of the plane can be found by obtaining the nearest points on two perpendicular lines in the plane, and adding their vectors.

Now the axes of co-ordinates do not enjoy any special privileges. All we know about the vectors $(1, 0, 0)$ and $(0, 1, 0)$ is that they are perpendicular, of unit length, and lie in the plane we are interested in. We expect the same result for any vectors with such properties in relation to any plane through the origin.

This can be verified directly. Let p and q be two perpendicular, unit vectors sprouting out from the origin. Any point in their plane is of the form $sp + tq$. The distance of the point given by this vector from U is d where

$$d^2 = \|u - sp - tq\|^2$$
$$= (u - sp - tq) \cdot (u - sp - tq)$$
$$= (u \cdot u) + s^2(p \cdot p) + t^2(q \cdot q) - 2s(u \cdot p) - 2t(u \cdot q) + 2st(p \cdot q)$$
$$= (u \cdot u) + s^2 + t^2 - 2s(u \cdot p) - 2t(u \cdot q).$$

The simplification in the last line is due to the fact that $(p \cdot p) = (q \cdot q) = 1$ since the vectors are of unit length, and $(p \cdot q) = 0$ since

they are perpendicular. The minimum occurs for $s = (u \cdot p)$ and
$t = (u \cdot q)$. These are precisely the values we would have obtained, if
we had sought separately for the point sp nearest to U and the point
tq nearest to U.

Again the connection between nearness and perpendicularity can
be verified. The vector joining the point $sp + tq$ to U is $u - sp - tq$. It
is perpendicular to p, since $(u - sp - tq) \cdot p = (u \cdot p) - s(p \cdot p) = (u \cdot p) - s = 0$. Similarly this vector is perpendicular to q. Accordingly it is perpendicular to every vector in the plane of p and q.

Exercises

1. Let $O = (0, 0, 0)$, $A = (0, 1, 1)$, $B = (1, 0, 1)$, $C = (1, 1, 0)$. Show that any
vector in the plane OBC is of the form $(s + t, t, s)$. What condition must s
and t satisfy if this vector is to be perpendicular to OB? Find the point, M,
of the plane OBC that is nearest to A. Calculate the distances $\|OM\|$ and
$\|AM\|$, and show that these agree with the values that could be deduced,
without the use of co-ordinates, from the geometry of the regular tetrahed-
ron, $OABC$, with side $\sqrt{2}$.

2. Let $A = (3, 4, 5)$ and $B = (5, 4, 3)$. Show that C, the point on the line
$x = y = z$ nearest to A, is also the point of that line nearest to B. Describe in
Euclid's language the figure formed by the points $O, A, B, C,$ and their
joins.

3. Let p_1, p_2, and p_3 be three mutually perpendicular vectors of unit length.
Let $v = c_1 p_1 + c_2 p_2 + c_3 p_3$. (The numbers c_1, c_2, c_3 are the co-ordinates of v
for axes in the directions of p_1, p_2 and p_3.) Find expressions giving c_1, c_2 and
c_3 in terms of scalar products of the four vectors.

Verify that $p_1 = (\frac{6}{7}, \frac{3}{7}, \frac{2}{7})$, $p_2 = (\frac{2}{7}, -\frac{6}{7}, \frac{3}{7})$ and $p_3 = (\frac{3}{7}, -\frac{2}{7}, -\frac{6}{7})$ are perpendicular
and of unit length. Find the co-ordinates, c_r, for $v = (5, 1, 1)$ and show that
these are the same as the co-ordinates of v in the original system.

About what line would the original axes of co-ordinates have to be
rotated to bring them into coincidence with p_1, p_2 and p_3?

8.2. Space of n dimensions

If the deductions in section 8.1 are examined, it will be found that
they do not depend on the space being of three dimensions. The
basic definitions are in terms of summations, and it does not matter
whether the sums are from 1 to 3 or from 1 to n.

In space of n dimensions the order of logical development is
usually slightly different. If we define the distance, $d(u, v)$, by

$$d(u, v)^2 = \sum (v_r - u_r)^2,$$

and the scalar product as $\sum u_r v_r$, we can then define the perpendicularily of two vectors as meaning that their scalar product is 0, and deduce Pythagoras' theorem from this, instead of the other way round. As we have seen, the further results can all be proved in terms of scalar products, without any use of summation signs. The algebraic properties of scalar products needed for this purpose are easily verified from the definition, $(u, v) = \sum u_r v_r$. It was mentioned earlier that from now on we would use (u, v) rather than $u \cdot v$.

When we consider projections in space of n dimensions, we do not have to confine ourselves to lines and planes. If we have k fixed vectors, $p_1, \ldots p_k$, all the points of the form $t_1 p_1 + \ldots + t_k p_k$, where $t_1, \ldots t_k$ are real numbers, form the subspace generated by the vectors $p_1, \ldots p_k$. We can ask what point of this subspace is nearest to a given vector u. If the vectors p_r are mutually perpendicular, we find, by the same method as before, that $t_r = (u, p_r)/(p_r, p_r)$. We should get the same number t_r if we looked for the point $t_r p_r$ nearest to u. As before, we can find the closest point in the subspace by combining the projections of u on the individual vectors p_r. If these vectors should happen to be of unit length, we would get the simpler formula, $t_r = (u, p_r)$.

The symbol $\mathscr{E}^n$ is used as an abbreviation for Euclidean space of n dimensions.

We do not have to look far for applications of $\mathscr{E}^n$, for in fact it was applied half a century before its existence was recognized. It was not until 1843–4 that Grassman and Cayley independently arrived at the idea of space with more than three dimensions. Already in 1795, Gauss used the method of least squares to determine the path of the planet Ceres. This method is used to extract the formula most likely to represent the law underlying a large number of observations that are subject to error. Suppose, for example, our observations give us n points, $(x_1, y_1) \ldots (x_n, y_n)$ that lie nearly but not exactly in line. If the underlying law is $y = mx + c$, the error in the rth observation is $y_r - mx_r - c$. The method of least squares consists in selecting those values of m and c that minimize the sum of the squares of the errors, $\sum_1^n (y_r - mx_r - c)^2$. Now this expression equals the square of the distance, in $\mathscr{E}^n$, between the points $(y_1, \ldots y_n)$ and $(mx_1 + c, \ldots mx_n + c)$. The second of these points may be written as $m(x_1, \ldots x_n) + c(1, 1, \ldots 1)$. That, it is a point in the plane of the two fixed vectors $(x_1, \ldots x_n)$ and

$(1, 1 \ldots 1)$. The values of m and c that make this point lie as close as possible to $(y_1, \ldots y_n)$ are accepted as the most probable values.

It should be noted that the two fixed vectors just mentioned will not, as a rule, be perpendicular, so the formula for m and c is a little more complicated than the one we have been using.

Gauss' problem was to find an unknown function. This however is not the only situation in which the least distance in $\mathscr{E}^n$ may be sought. It is also relevant to approximation theory, in which effective ways of calculating known functions are sought. Much computer time would often be required to calculate the values of a function from its usual analytic definition. Since the computer in any case obtains the values only to a certain number of decimal places, it is often possible to find some simpler formula that reproduces the values to the required degree of approximation. Such approximate formulas are used in all computers, from the most massive and expensive to the cheapest pocket computer capable of handling logarithms or trigonometric functions. An extensive collection of suitable approximations can be found in Hart.

That such approximations can be found by projection in $\mathscr{E}^n$ is shown in the following paragraph. The procedure described there is is not always to be recommended; it can be made more reliable and effective by refinements that will be introduced in section 7.4, when approximation by Chebyshev polynomials is explained.

As a very crude example, suppose we are given the sines of $0°$, $30°$, $60°$, $90°$ and wish to obtain a simple approximation to the sine function. Our data give us $\sin 30x°$ for $x = 0, 1, 2, 3$. As $\sin(-\theta) = -\sin\theta$, it will be good if we approximate using only odd powers of x. The approximation will then automatically work as well for negative values of x as for positive. It would be natural to consider $ax + bx^3$ as a possible simple approximation. As will be seen in a moment, it proves convenient to put the approximation in the form $sx + t(x^3 - 7x)$. We then have the following table.

x	0	1	2		3	p
$x^3 - 7x$	0	-6	-6		6	q
$\sin 30x°$	0	0.5	0.866 025		1	u

The four numbers in each row define a vector in $\mathscr{E}^4$. These are labelled p, q, u respectively. Our problem is to find the combination

$sp + tq$ nearest to u. The condition $p \perp q$ is satisfied. It was in order to obtain this that we chose $x^3 - 7x$ for our second ingredient rather than x^3. However p and q are not of unit length. We have $(p, p) = 14$, $(q, q) = 108$, $(u, p) = 5.232\,050$, $(u, q) = -2.196\,150$, from which it follows that

$$s = (u, p)/(p, p) = 0.373\,718$$

and

$$t = (u, q)/(q, q) = -0.020\,335.$$

After substituting these values and simplifying we obtain the approximation $0.516\,061x - 0.020\,335x^3$. The following table shows how this compares with the accurate values.

Angle	0°	30°	60°	90°
Sine	0	0.5	0.866 025	1
Approximation	0	0.495 726	0.869 444	0.999 145
Error	0	0.004 274	−0.003 419	0.000 855

The closeness of fit is not too bad, in view of the fact that the whole range, $-90°$ to $90°$, is being covered and that the method is extremely elementary. No appeal has been made to the infinite series for $\sin \theta$; only the values of $\sin 30°$ and $\sin 60°$, which can be obtained from the geometry of the equilateral triangle, together with the obvious results for $\sin 0°$ and $\sin 90°$, have been used. If this approximate formula was used to generate a table of sines, the largest error would occur near to $74°$ and would be of the order of -0.0065.

Exercises

1. Prove that the values of f and g, taken for the x-values $-2, -1, 0, 1, 2$, give perpendicular vectors if $f(x) \equiv 1$ and $g(x) \equiv x^2 - 2$. Find the values of a and b that make $a + b(x^2 - 2)$ the best approximation (in the sense of least squares) to $e^{-x^2/8}$ for the x-values listed above. Tabulate these, and compare them with the errors for the approximation $1 - (x^2/8)$, given by the beginning of the Taylor series for $e^{-x^2/8}$.

2. Verify that 1 and $x^2 - 3.5$ give perpendicular vectors when the x-values 0, 1, 2, 3 are successively substituted. Find the least squares approximation of $a + b(x^2 - 3.5)$ to $\cos 30x°$ for these values.

If you differentiated the least squares approximation to the sine function, would you expect to get the approximation to the cosine function?

3. State problems equivalent to (1) and (2) above as questions about points and planes in Euclidean spaces of suitable numbers of dimensions.

8.3. Space of infinite dimensions

The considerations above lead quite naturally to a second stage of generalization. We have just fitted an approximate formula to the sine function at the points 0°, 30°, 60° and 90°. If we had the data, we could fit the approximation to a larger number of points; the method does not place any restriction on the number of data. If $f(x)$ was to be an approximation to sin x°, based on data for every whole number of degrees, we would consider the sum

$$\sum_{n=0}^{n=90} [\sin n° - f(n)]^2.$$

A sum, taken at such a large number of points must be closely related to an integral, and this suggests the thought that if we replaced sums by integrals, we could get an approximation based on the values for *every* point of the interval.

Degrees and integration do not go well together, so let us consider $f(\theta)$, which is to be an approximation to sin θ, where θ is in radian measure. To estimate the closeness of fit we use the integral

$$\int_0^{\pi/2} [\sin \theta - f(\theta)]^2 \, d\theta.$$

If f is to be a function of specified type, say given by a polynomial of specified low degree, we shall choose the coefficients so as to minimize this integral.

On the analogy of the work in section 7.2, we may assume $f(\theta) = s\theta + t(\theta^3 - k\theta)$. The most convenient value for k will appear in the course of the work. Accordingly we seek to minimize

$$\int_0^{\pi/2} [\sin \theta - s\theta - t(\theta^3 - k\theta)]^2 \, d\theta.$$

The appearance of the integral here may suggest that we are now dealing with a much harder problem, perhaps one involving the calculus of variations. Actually, we are not. The variables are s and t, and the expression to be minimized is a quadratic in them. The coefficients happen to be given by certain definite integrals. For instance, s^2 occurs with the coefficient $\int_0^{\pi/2} \theta^2 \, d\theta$, which is $\pi^3/24$, approximately 1.291 928.

At the end of section 8.1, a quadratic in s and t was minimized. There the calculation was simplified by the fact that, owing to $p \perp q$, the coefficient of st was 0. We can arrange for this coefficient to be 0 here too, by choosing a suitable value for k. The coefficient is $2\int_0^{\pi/2} \theta(\theta^3 - k\theta) \, d\theta$, which is $(\pi^5/160) - k(\pi^3/24)$. This can be made 0 by choosing $k = 3\pi^2/20$, which we suppose done. The expression to be minimized then is found to be

$$0.785\,398 + 1.291\,928s^2 - 2s + 0.539\,338t^2 + 0.156\,475t.$$

The terms in s are minimized by $s = 0.774\,037$, those in t by $t = -0.145\,062$. Substituting these in the expression for $f(\theta)$ we find $f(\theta) = 0.998\,792\theta - 0.145\,061\theta^3$. As $30x°$ is the same as $x(\pi/6)$ radians, the substitution $\theta = \pi x/6$ gives us a result that can be compared with our earlier approximation. It is $0.517\,730x - 0.020\,823x^3$, which is not very different from our earlier result, $0.516\,061x - 0.020\,335x^3$. This is not surprising. There must be some limitation on how closely a cubic curve can fit a sine curve. Even if we have a very elaborate procedure for selecting the best cubic, we cannot expect it to make a great improvement in the closeness of fit.

The significance of the work just done therefore does not lie in the improvement of particular approximations. Rather it lies in the fact that we have successfully applied to sine and polynomial functions a procedure originally developed for vectors in $\mathscr{E}^3$ or $\mathscr{E}^n$. We are in fact in a position to define the distance between two functions, to speak of one function being perpendicular to another, to consider projections onto subspaces generated by one or more functions. The theory follows the lines of our earlier remarks; we simply replace summation by integration, taken over any convenient, fixed, finite interval $[a, b]$.

We define the norm of a function, f, by

$$\|f\|^2 = \int [f(x)]^2 \, dx,$$

and the scalar product of two functions, f and g, by $(f, g) = \int f(x)g(x) \, dx$. The functions will be called perpendicular when $(f, g) = 0$. It seems customary to use the word 'orthogonal', a synonym of 'perpendicular', when speaking of functions, but the meaning is exactly the same. The distance of f from g is $\|f - g\|$. As (f, g) depends linearly on f when g is fixed, and linearly on g when f is fixed, its algebraic properties are easily deduced, and we can carry our experience with dot products over to the manipulation of (f, g). In fact, those arguments in section 8.1 which are in terms of scalar products can be carried through without stopping to enquire whether the u and v in (u, v) are vectors in $\mathscr{E}^3$, vectors in $\mathscr{E}^n$, or functions $[a, b] \to \mathbb{R}$.

The symbol $\mathscr{L}_2[a, b]$ will be used to indicate the space consisting of real-valued functions, defined on $[a, b]$, with

$$\|f\|^2 = \int_a^b [f(x)]^2 \, dx.$$

We have come to this space without any discussion of the theoretical problems involved. These include the question of what functions, f, are acceptable, and involve the modern theory of integration. Some notes on these matters will be given in section 8.5. We have not even proved that our definition satisfies the norm axioms, although in fact it does do so. In fact, so far we have not been discussing the nature of the space $\mathscr{L}_2[a, b]$ at all, but have been concentrating on the fairly elementary problem of approximating, in the sense of least squares, to a function f by an expression $\sum t_r p_r$. In the applications of this we have in mind, the functions, $p_1, \ldots p_n$, will be continuous. The function, f, may not be, but it will be capable of integration by elementary methods. It may, for instance, be defined by a finite number of different formulas, each applying to some part of the interval $[a, b]$. In each of these parts, the function will be continuous.

Exercises

1. What, in terms of (u, u), (u, v) and (v, v), is the distance of tv from u, when t is chosen to make this a minimum?

2. Prove that, in $\mathscr{L}_2[0, \pi/2]$, the best approximation to $\sin x$ of the form mx is given by $m = 24/\pi^3$. What is the $\mathscr{L}_2$ distance of the approximation from the sine function?

3. Show that the algebraic arguments of section 8.3 are justified when (u, v) is defined as $\int_0^{\pi/2} u(x)v(x) \, dx$.

4. Let $f(x) \equiv a + bx^2$, $g(x) \equiv cx + kx^3$. Find the angle between the vectors that represent f and g in $\mathscr{L}_2[-1, 1]$.

5. Let $f(x) \equiv 0$, $g(x) \equiv 2$, $h(x) \equiv 1 + 3x$. Find the distances $d(f, g)$, $d(f, h)$, $d(g, h)$ in $\mathcal{L}_2[-1, 1]$. What geometrical figure is formed by the points representing f, g, h in this space? What value does this geometrical representation suggest for the scalar product (g, h)? Test the validity of this suggestion by calculating (g, h) directly as an integral.

6. Let $f_n(x) \equiv x^n$. Find $\|f_n\|$, (a) if f_n is regarded as an element of $\mathcal{L}_2[0, 1]$, (b) if f_n is regarded as an element of $\mathscr{C}[0, 1]$. In each case, discuss whether f_n tends to a limit as $n \to \infty$, and state the limit if it does.

7. Do as in question 6, with $f_n(x) \equiv n^{\frac{1}{2}} e^{-nx}$.

8. Let $f(x) \equiv a$ and $g(x) \equiv \cos x$. Find what value of a makes the distance of f from g in $\mathcal{L}_2[-\pi/2, \pi/2]$ a minimum, and determine this minimum distance. For this value of a, what is the distance of f from g in $\mathscr{C}[-\pi/2, \pi/2]$? What value of a would make f nearest to g in $\mathscr{C}[-\pi/2, \pi/2]$, and how far apart would they then be in that space?

9. Let $f(x) \equiv ax$ and $g(x) \equiv \sin x$. Find what value of a makes $d(f, g)$ in $\mathcal{L}_2[0, \pi/3]$ a minimum, and find this minimum distance.

A well-known crude approximation to $\sin x°$ is $x/60$, which corresponds to $a = 3/\pi$ for radian measure.

The value of a that makes f closest to g in $\mathscr{C}[0, \pi/3]$ is 0.869 647 167.

Make tables showing the errors of approximation, $ax - \sin x$, for the three values of a referred to in the paragraphs above, and compare them.

8.4. Fourier series and orthogonal functions

Two functions, p and q, defined on $[a, b]$, as we have seen, are called orthogonal if

$$0 = (p, q) = \int_a^b p(x)q(x)\, dx.$$

A famous example is given by $p(x) = \sin mx$, $q(x) = \sin nx$, on the interval 0, where m and n are whole numbers, and $m \neq n$. The proof follows immediately from the trigonometrical identity

$$2 \sin mx \sin nx = \cos(m - n)x - \cos(m + n)x.$$

The right-hand side, after integration from 0 to π, gives 0.

When $p_1, \ldots p_n$ are mutually orthogonal, the best approximation to f by $\sum c_r p_r$ has $c_r = (f, p_r)/(p_r, p_r)$. If we take $p_r(x) = \sin rx$, on $[0, \pi]$, the functions $p_1, \ldots p_n$ will be orthogonal. As

$$(p_r, p_r) = \int_0^\pi (\sin rx)^2\, dx = \pi/2,$$

we shall have

$$c_r = (2/\pi)(f, p_r) = (2/\pi) \int_0^\pi f(x) \sin rx \, dx. \tag{1}$$

The numbers c_r are known as Fourier coefficients of f.

The approach used here is far different from the historical development. It may encourage students who find that it takes some time to understand a new mathematical idea to realize that it took mathematicians of genius a century and three-quarters to disentangle the issues involved in this topic. The term 'orthogonal functions' seems to have been used first by Erhardt Schmidt in 1907. The term, Fourier series, for a series involving sines, cosines, or both, goes back to a publication by Fourier in 1811. Results such as equation (1) of this section were known to Clairaut in 1757, and the topic originated in investigations of musical vibrations by Daniel Bernouilli in 1730. It was the subject of heated debate throughout the eighteenth century.

The method of obtaining the series in those days was quite different. An infinite series was assumed, of the form

$$f(x) = c_1 \sin x + c_2 \sin 2x + c_3 \sin 3x + \ldots. \tag{2}$$

The equation was then multiplied by $\sin rx$, and the result integrated from 0 to π. On the right-hand side, all the integrals gave 0, except that containing $c_r(\sin rx)^2$, which gave $\pi/2$, and led to

$$\int_0^\pi f(x) \sin rx \, dx = c_r(\pi/2). \tag{3}$$

The result is equivalent to (1). This method assumes that a sine series giving $f(x)$ does exist, and that term by term integration of an infinite series is justified. It was only realized after 1847, when the concept of uniform convergence was discovered, that this was not always permissible.

In equation (2), the equals sign presumably implies that the series will sum to $f(x)$. In modern treatments, this is often replaced by the sign, $\sim$. This is to be interpreted in the following way. It means that, if we start with the function f, and calculate the numbers c_r given by equation (1), these will be the same as the coefficients on the right-hand side of equation (2). Whether the resulting series will converge, and if so, whether it will converge to $f(x)$, is a matter for subsequent investigation.

The thing that was not suspected at all in the early development of Fourier series was that the numbers, c_r, could be regarded as co-ordinates in a space of infinitely many dimensions, and the sine functions as perpendicular vectors, of length $\sqrt{(\pi/2)}$, in this space. In an orthogonal co-ordinate system, we usually take the basis vectors of unit length. This suggests that we bring in functions, q_r, with $q_r(x) = \sqrt{(2/\pi)} \sin rx$. The projections of f on these vectors will be $k_r q_r$, where

$$k_r = (f, q_r) = \int_0^\pi f(x) \cdot \sqrt{(2/\pi)} \cdot \sin rx \, dx,$$

since $(q_r, q_r) = 1$. The vectors q_r are mutually perpendicular and of unit length. Such a collection of vectors is called an orthonormal system (a shortened form of 'orthogonal, normalized' system). The abbreviation o.n.s. is sometimes used.

Our geometrical picture now begins to work for us. Fourier series have played a very important part in the history of mathematics. One could be forgiven for thinking they constituted a unique phenomenon. Our vector picture immediately shows that it is not so. There are infinitely many ways of putting a set of perpendicular, unit vectors into a Euclidean space. There must be the freedom to do the same in the space $\mathscr{L}_2[a, b]$. Changing from one such system to another must resemble rotation of axes.

In section 7.2 we found it convenient to consider x and $x^3 - 7x$ rather than x and x^3; similarly in section 7.3 we considered θ and $\theta^3 - k\theta$. The advantage lay in the fact that these were pairs of perpendicular vectors, which we manufactured by forming a suitable combination of two vectors already given, such as x and x^3 in the first example. There is in fact a procedure for obtaining an orthonormal system equivalent to any given set of vectors, in the sense that the same subspace is generated.

In Figure 35, we see a horizontal vector, u, and an inclined vector, v. The plane of the paper is generated by u and v. In this plane,

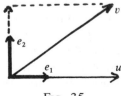

FIG. 35

there is obviously a vertical vector, which can be expressed as a combination of u and v. The plane generated by u and such a vertical vector is the same as that generated by u and v.

The algebraic procedure is the following. Dividing u by its length, $\|u\|$, we reach a vector of unit length, $e_1 = u/\|u\|$. We now consider $v + ke_1$. This will be perpendicular to

$$e_1 \quad \text{if} \quad 0 = (v + ke_1, e_1) = (v, e_1) + k(e_1, e_1) = (v, e_1) + k,$$

since $\|e\| = 1$. So we take $k = -(v, e_1)$. In fact, the required vertical vector is obtained by subtracting from v its projection on the horizontal. Having obtained the vertical vector, we take a unit vector in its direction, and so have e_2. The system $\{e_1, e_2\}$ is orthonormal.

It is possible to continue. Suppose there is a third vector, w, not lying in the plane of the paper. We can construct a vector e_3, perpendicular to the plane of the paper, and of unit length. First we consider $w + c_1e_1 + c_2e_2$. By forming scalar products, we see that this vector will be perpendicular to e_1 if $c_1 = -(w, e_1)$ and to e_2 if $c_2 = -(w, e_2)$. Taking these values, we have a vector perpendicular to the plane of the paper. Dividing this vector by its length, we obtain the vector of unit length, e_3. The system $\{e_1, e_2, e_3\}$ is equivalent to u, v, w and is orthonormal.

This procedure can be extended to deal with any number of vectors. It is known as the Gram–Schmidt procedure, after two mathematicians who worked in this field.

If we apply this procedure to $1, x, x^2, x^3, \ldots$ on the interval $[-1, 1]$, we obtain a sequence of orthogonal polynomials. The nth polynomial is in fact proportional to $(d/dx)^n(x^2 - 1)^n$. These polynomials are closely related to the Legendre polynomials, usually denoted by $P_n(x)$, introduced in 1784 and of importance in mathematical physics. The difference lies only in the normalization. The Legendre polynomials have $P_n(1) = 1$. To normalize them in our sense, $P_n(x)$ has to be multiplied by a factor $\sqrt{(n + 0.5)}$.

The Legendre polynomials provide a simple example of orthogonal functions, but they do not appear to be very convenient for numerical work (see Clenshaw 3, chapter 3, section 4). In contrast to this, another sequence of polynomials, the Chebyshev polynomials, are of great importance for numerical analysis. These were introduced by Chebyshev in 1853, in a paper on the best polynomial approximations to given functions, and have proved of great value ever since.

The Chebyshev polynomials can be defined in a very elementary way. If n is a whole number, $\cos n\theta$ is a polynomial of degree n in $\cos \theta$. If $x = \cos \theta$, the Chebyshev polynomial $T_n(x) = \cos n\theta$. For example, $\cos 2\theta = 2(\cos \theta)^2 - 1$, so $T_2(x) = 2x^2 - 1$. The letter T is used for these polynomials because Tchebyshev and Tschebyscheff are among the many ways of writing this name in our alphabet. The trigonometric identity

$$\cos n\theta + \cos(n-2)\theta = 2 \cos(n-1)\theta \cos \theta$$

immediately gives the recurrence relation

$$T_n(x) = 2xT_{n-1}(x) - T_{n-2}(x).$$

As $T_0(x) = 1$ and $T_1(x) = x$, the value of $T_n(x)$ can be calculated.

In $\mathscr{L}_2[0, \pi]$ the functions given by $\cos n\theta$ are orthogonal. It can be verified that, if m and n are whole numbers, with $m \neq n$, then

$$0 = \int_0^\pi \cos m\theta \cos n\theta \, d\theta.$$

If we put $x = \cos \theta$, this equation becomes

$$0 = \int_{-1}^1 \frac{T_m(x)T_n(x)}{\sqrt{(1-x^2)}} \, dx.$$

Here we have yet another extension of the idea of orthogonality. In section 8.3 we thought of integrals arising from sums as more and more points were taken on an interval. At that time, we imagined these points at each stage evenly spaced along the interval. But we are not forced to do that. We could consider least square problems in which points clustered more densely in some parts of the interval than in others. In that case, the sum defining (f, g) would in the limit approach $\int_a^b f(x)g(x)w(x) \, dx$, where w is known as a weight function, and indicates how the density of the points varies from place to place. It will be seen that for the Chebyshev polynomials, the factor, $w(x) = 1/\sqrt{(1-x^2)}$, puts extreme weight on what happens near the ends of the interval, $[-1, 1]$. In many cases this proves a very wise thing to do. A very readable account of the tendency of least squares problems to become ill-conditioned, and of the ways in which Chebyshev polynomials help us to avoid this, will be found in Hayes, chapter 5. Much information on Chebyshev polynomials and tables for the approximation of functions by them will be found in Clenshaw (1). L. Fox also gives very useful information in chapters 4, 5 and 6 of Handscomb.

In approximation work our usual aim is to get a table accurate to a certain number of decimals. We are accordingly concerned about the maximum error that occurs in the table; we want this to be as small as possible. The polynomial that does best in this sense to fit a given function is a prescribed interval is known as the *minimax polynomial*. This is the polynomial we would most like to know. Unfortunately, there is no obvious algorithm for obtaining it. There is a simple and reliable procedure for combining Chebyshev polynomials to fit a function, but of course the polynomial that results from this procedure is not the minimax polynomial. Fortunately, it is a good substitute. In 1964 Clenshaw arrived at the belief that 'in all practical circumstances the truncated Chebyshev series is almost as good as the minimax polynomial. For example, in proceeding from the former to the latter, provided only that $f(x)$ is continuous, we cannot gain one extra decimal figure of accuracy unless the degree of the polynomial exceeds 400.' He proved this for certain cases and conjectured that it was true generally. This was later proved to be the case in a joint paper by Binh Lam and David Elliott. (See Clenshaw (2, 3); Lam and Elliott; also problem 13 in chapter 4, section 6, of Cheney.)

Not all curves are suited for polynomial approximation. The example now to be considered shows two things—the danger of

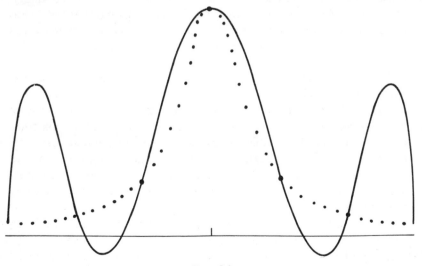

Fig. 36

using a method without a knowledge of its theoretical background and the inherent limitations in polynomial approximation. In Figure 36, the dotted curve represents a function, given in the interval $-1, 1$, which we wish to approximate. A method, that appears reasonable, would be to take a number of points on the curve, and choose a polynomial whose graph goes through them. The continuous curve in Figure 36 is obtained in this way, the x-values chosen being -1, $-2/3$, $-1/3$, 0, $1/3$, $2/3$, 1. It will be seen that our approximating curve has a high peak at each end which does not correspond to anything on the original curve. It might be thought that this is because we have not taken enough points on the original curve. The reverse is true. As the number of evenly spaced points increases indefinitely, the height of the peaks tends to infinity. This effect is known as the Runge phenomenon. The theory of it will be found in Todd, section 3.6.

We do not get any such spectacular disaster if we approximate by Chebyshev polynomials. It is convenient to write the series in the form $(c_0/2) + \sum_1^\infty c_n T_n(x)$. The reason for this is that the Chebyshev series is a disguised Fourier series, and $\int_0^\pi (\cos n\theta)^2 \, d\theta$ is $\pi/2$ when $n \neq 0$, but π when $n = 0$. The example we are considering is exceptional, in that there is a simple formula for the coefficient c_n, namely $c_n = c_0 r^n$, with $c_0 = 0.392\,232\,270$ and $r = -0.672\,078\,439$. As $T_n(x) = \cos n\theta$, the value of $T_n(x)$ is always from -1 to $+1$. Thus the coefficients themselves give an indication of whether the series is converging and how fast. As $|r| < 1$, the coefficients c_n form a convergent geometrical progression, but c_{20} is still as large as $0.000\,139\,121$. Thus a computation involving 21 terms is still far from giving 6 place accuracy. Polynomial approximation in fact is not the best method here. Actually, the dotted curve has a quite simple equation, $y = 1/(1 + 25x^2)$. Of course, if we knew this from the start, we would not be seeking a polynomial approximation. If we came across a curve that did not coincide with this one but resembled it, a sensible procedure would be to seek a rational approximation, or apply some transformation before approximating with polynomials.

The unsuitability of this curve for polynomial approximation could be seen from the fact that the derivatives $f'(x)$, $f''(x) \ldots$ take high values in $[-1, 1]$. This in turn is due to the fact that $1/(1 + 25x^2)$ becomes infinite for $x = \pm 0.2\sqrt{(-1)}$. These values are complex but they affect polynomial approximation. The theory of this is discussed by D. F. Mayers in chapter 3, sections 7 and 8, of Handscomb. Other chapters of this book also provide excellent summaries of theory and practice in numerical analysis.

Exercises

1. Show that for $f(x) \equiv \pi^2 x - x^3$ with the interval $[0, \pi]$, equation (1) gives the Fourier coefficients, $c_r = 12(-1)^{r+1}/r^3$. Use these values to calculate

$$g(x) = \sum_1^{10} c_r \sin rx$$

for $x = n\pi/4$, with n taking whole number values from -4 to 12 inclusive. Sketch a rough graph of g for $-\pi \le x \le 3\pi$, and compare this with the graph of f.

2. (The sawtooth function.) Let $f(x) = x$ for $0 \le x \le \pi/2$ and $f(x) = \pi - x$ for $\pi/2 < x \le \pi$. Show that the corresponding Fourier series is

$$g(x) = (4/\pi)[\sin x - (1/9)\sin 3x + (1/25)\sin 5x \ldots]$$

with the squares of the odd numbers appearing as denominators.

Does the series for $g(x)$ converge (a) absolutely, (b) uniformly? Investigate the graph of g, with x taking all real values.

Sketch the graph of the derivative f' in $[0, \pi]$. Would you expect a Fourier series for f' to converge uniformly?

3. Show that the Fourier series $(4/\pi) \sum_0^\infty [1/(2r+1)^2]\cos(2r+1)\theta$ arises from $f(\theta) \equiv (\pi/2) - \theta$ on $[0, \pi]$. By putting $x = \cos \theta$ obtain the series of Chebyshev polynomials that corresponds to $\sin^{-1} x$.

Tabulate the errors produced when $\sin^{-1} x$ is approximated (a) by the terms up to x^5 in its Taylor series, (b) by the Chebyshev series up to the term in $T_5(x)$. Observe the contrast in behaviour between these.

8.5. The theoretical background

In this section we consider certain things of which numerical analysts should be aware, even if these things are not immediately used in any particular algorithm, for we have now reached a stage in mathematics where subtleties play a significant role. The first university course in analysis often strikes students as a time-wasting exercise in pedantry. Such immense care is taken to prove such obvious results. At school students had often done calculus with the help of nothing more than robust common sense, and it seemed to work very well. The reason why it worked so well was that it stopped short of Fourier series. Experience in school usually does not go further than power series, such as those for e^x, $\sin x$, $\ln(1+x)$, and so forth. Within their region of convergence, power series converge uniformly and absolutely, which means that you get correct results by doing almost anything a sane person might consider. It is not so with Fourier series. They behave in most unexpected ways, and a large part of the development of classical analysis was forced by this fact.

Consider for example the series $\sum c_n \sin nx$ that arises from the function f with $f(x) \equiv 1$ in $[0, \pi]$. Putting 1 for $f(x)$ in the formula

$$c_n = (2/\pi)\int_0^\pi f(x)\sin nx \, dx,$$

we easily obtain the series

$$\phi(x) = (4/\pi)[\sin x + (1/3)\sin 3x + (1/5)\sin 5x + \ldots]$$

For $0 < x < \pi$, the series does in fact converge to 1, from which we set out. As $\sin n(-x) = -\sin nx$, for $-\pi < x < 0$, $\phi(x) = -1$. If x is $-\pi$, 0 or π, $\phi(x) = 0$. The function ϕ is periodic with period 2π, so its graph repeats itself indefinitely, like a wallpaper. The graph is shown in Figure 37. On account of the graph's appearance, the function ϕ is sometimes called 'the parapet function'.

Such behaviour disturbed eighteenth-century mathematicians very much. If you are given a small part of the graph of a polynomial or an analytic function, you can (in theory) deduce the entire graph from this small piece. Fourier series are quite different; their graphs can be a collection of unrelated bits and pieces. This is true whether the series consists of sines alone, cosines alone, or both.

We have seen that the series of Chebyshev polynomials, $\sum c_n T_n(x)$, can be converted by the substitution $x = \cos\theta$ into the Fourier series, $\sum c_n \cos n\theta$. This means that $\sum c_n T_n(x)$ can show the type of behaviour just described. That is something a power series

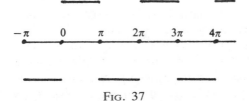

$$-\pi \quad 0 \quad \pi \quad 2\pi \quad 3\pi \quad 4\pi$$

FIG. 37

cannot do, so here already we have a subtlety; a series of Chebyshev polynomials, although it is made up of powers of x, is not a power series! The rearrangement of the powers of x has made an essential difference both to its purely mathematical properties and to its behaviour in regard to computing operations. It is possible, for instance, for a Chebyshev series to be well-conditioned while a corresponding power series is ill-conditioned. (See Hayes, chapter 5.)

Obviously some complication must be present when the sum of a series of continuous functions, such as sines, turns out to be discontinuous. What happens is that the rate of convergence slows down as the discontinuity is approached. For instance, in order to get values of the parapet function, ϕ, lying between 0.99 and 1.01, it is necessary to take about 40 terms for $x = 0.1$, about 95 for $x = 0.01$ and about 950 for $x = 0.001$.

The Chebyshev series

$$(4/\pi)[T_1(x) - (1/3)T_3(x) + (1/5)T_5(x) - (1/7)T_7(x)\ldots]$$

sums to 1 for $0 < x < 1$ and to -1 for $-1 < x < 0$. This can be verified in a very rough manner by summing the first 30 terms of the series, provided values not too close to the discontinuity at $x = 0$ are taken. The values of $T_n(x)$, for any particular x, can be found by the recurrence relation corresponding to the trigonometric identity $\cos(n+2)\theta + \cos(n-2)\theta = 2\cos n\theta \cos 2\theta$. The rounding-off errors produced by this recurrence procedure can be estimated by comparing $T_n(x)$ with $\cos n\theta$, where $\cos \theta = x$. The complexity of the behaviour of Fourier series has forced mathematicians to develop the theory of integration. The first step in this direction was taken by Riemann in 1854. Riemann had come across a function, represented by a Fourier series, the graph of which could not be drawn because it consisted of disconnected points. In any interval, however short, there were an infinity of discontinuities. Yet there were reasons for believing that the integral, the area under this graph, had a meaning.

Riemann's example was of the following type. We begin the construction by considering $F(x)$ where

$$F(x) = 2[\sin x - (1/2)\sin 2x + (1/3)\sin 3x - (1/4)\sin 4x\ldots].$$

For $-\pi < x < \pi$, $F(x) = x$. Now F is periodic with period 2π, so between $x = \pi$ and $x = 3\pi$, $F(x)$ again climbs from -1 to 1. This means $F(x)$ has a discontinuity at $x = \pi$, and similarly it has discontinuities at 3π, 5π, 7π and so on. Now consider $F(2x)$; it will have discontinuities at $\pi/2$, $3\pi/2$, $5\pi/2$, ... Similarly $F(4x)$ will show discontinuities at $\pi/4$, $3\pi/4$, $5\pi/4$, $7\pi/4$ ... Now let

$$G(x) = F(x) + (1/2)F(2x) + (1/4)F(4x) + (1/8)F(8x) + \ldots$$

This will show discontinuities at every point of the form $x = m\pi/2^n$, where m and n are whole numbers. It will be an ungraphable function of the type described at the outset. Yet each term of the series for $G(x)$ is expressible as a Fourier series, and the sum of all these Fourier series looks like being a Fourier series. As the coefficients of the Fourier series are found by equation (1), involving integration, it seems that F, in spite of its many discontinuities, should possess an integral.

But even Riemann's theory was not sufficient for the needs of Fourier series. His theory is equivalent to the following. To estimate $\int_a^b f(x)\, dx$, divide $[a, b]$ into N intervals of lengths $k_1, \ldots k_N$. Let the maximum values of $f(x)$ in these intervals be $M_1, \ldots M_N$ and the minimum values $m_1, \ldots m_N$. (Strictly, we should say supremum and infimum.) We take $\sum M_r k_r$ as an overestimate and $\sum m_r k_r$ as an underestimate of the integral. If these two estimates tend to a common limit as we take increasingly fine subdivisions of $[a, b]$, this common limit defines the values of the integral, and f is said to be Riemann-integrable.

Now consider Figure 38, in which certain rectangles are seen standing on the interval $[0, 1]$. The rectangles are all supposed to be of height 1. For clarity it has been necessary to enlarge the horizontal scale. The construction can be imagined as follows. We begin with a rectangular piece of black paper, of height 1 and base length $1/2$. We cut it into two equal pieces by a vertical line, so we now have two pieces, each of base length $1/4$. One of these we paste to the graph paper, so that it is centred over the point with $x = 1/2$. It thus stands on the interval $[3/8, 5/8]$. There are now two unoccupied intervals, $[0, 3/8]$ and $[5/8, 1]$. We have material in hand, of width $1/4$. We conserve half of this for later

Fɪɢ. 38

operations (as we shall do at every stage of the process), and share half of it fairly between the unoccupied intervals. Thus we now centre pieces of width 1/16 over the points 3/16 and 13/16. We continue with this policy. At each stage we use only half the material in hand; we divide this evenly into as many pieces as there are unoccupied intervals at that stage, and place a piece centrally in each such interval. Thus, at the third stage, four pieces of width 1/64 will be centred over the dotted lines in the figure. This process is supposed to go on for ever.

We now regard this set of rectangles as defining the graph of a function. That is to say, we take $\phi(x) = 1$ if the point x gets covered by a rectangle at any stage of the process, and $\phi(x) = 0$ otherwise. It will also be convenient to define functions corresponding to the stages of this construction. Thus $\phi_k(x) = 1$ if x has been covered by a rectangle when the kth stage of the construction is complete, and $\phi_k(x) = 0$ otherwise.

Two things will now be asserted about the function ϕ. First, although ϕ is defined in a rather peculiar way, we ought to accept it as a member of the space $\mathscr{L}_2[0, 1]$. Second, this means that we have to extend our definition of integration, for $\int_0^1 [\phi(x)]^2 \, dx$ is not meaningful by Riemann's definition.

As to the first, it is inconvenient to work in a space that is not complete. That means that we have sequences which pass the test for convergence (Cauchy sequences) but we reject the function to which they tend because we do not like the look of it. Now in the $\mathscr{L}_2$ metric, the functions ϕ_k approach ϕ, and there are continuous functions (easily constructed) that approach each ϕ_k. If ϕ is not admitted, our space will be incomplete.

We now consider the Riemann definition of integral in relation to ϕ. However many intervals (finite in number) we divide [0, 1] into, we will always find in each interval a point covered by a rectangle. That is, in each interval, there will be a point with $\phi(x) = 1$, so the maximum value $M_r = 1$. Accordingly our overestimate of the integral will always be 1. We can get underestimates close to 1/2 by taking minimum values, but there is no way of closing this gap between 1/2 and 1. Thus the Riemann definition of integral breaks down.

Since $\phi(x)$ only takes the values 0 and 1, $[\phi(x)]^2 \equiv \phi(x)$. So the Riemann definition does not allow us to cope with the integral of $[\phi(x)]^2$, which we need to calculate the norm of ϕ.

Now a commonsense argument suggests that $\int_0^1 \phi(x) \, dx$ ought to exist and have the value 1/2. For this integral corresponds to the area of the black rectangles in our construction. As these were all made from a rectangle of height 1 and width 1/2, their total area can hardly exceed 1/2. On the other hand, there is no material to spare, so we have no reason to believe that the total area can be less than 1/2.

The difference between this argument and Riemann's is that we have covered the black regions with an *infinity* of rectangles. In Riemann's theory, we are only allowed to form an overestimate of the area of a region by covering it with a *finite* number of rectangles. This is the crucial change that led to the second stage in the development of integration theory. In 1898 Borel suggested that infinite sums might be permitted in the theory, and in various papers and a celebrated dissertation of 1902, Lebesgue worked out the application of this idea to the theory of lengths, areas and integrals.

Throughout the ages there has been a continuing oscillation in the attitude of mathematicians to the idea of infinity. The ancient Greeks would have been shocked by the construction above for the graph of ϕ, in which the placing of an infinity of rectangles was regarded as having been completed. They held that infinity was concerned only with possibilities, never with something actually finished. They would not have said, as we do, that there are infinitely many prime numbers, but only that, however many prime numbers you may name, others remain not named. In the seventeenth and eighteenth centuries, most mathematicians were eagerly applying calculus to a host of practical and theoretical questions, and in the main had a very happy-go-lucky attitude to infinity and infinitesimals. The reaction to this came in the nineteenth century, with the development of rigour in analysis. There was a horror of infinity. Infinity was not a number. 'Tends to infinity' was a convenient phrase, an abbreviation for a phrase that only involved finite numbers and that the Greeks would have approved of. This made things very difficult for Cantor, with his ideas of completed infinities, some of which were larger than others. It is quite acceptable to-day to say that the entire plane has an area of $+\infty$, to suppose that infinite constructions of point sets have been completed, and to build theories of real, practical use, such as Lebesgue integration, on the basis of this faith. If civilisations capable of supporting mathematics continue, two things appear almost certain—that practical methods, such as Lebesgue integration, will remain as a permanent possession of ours, and that the theories of infinity, used to justify such applications, will continue to fluctuate as the centuries pass.

Integrals are associated with areas. If we are dealing with a function f on some interval $[a, b]$, we can imagine a vertical line of length $|f(x)|$, drawn away from the x-axis, at each point x of the interval. If $f(x) > 0$, this line is drawn upward; if $f(x) < 0$, downward. The totality of these lines form a region. If A is the area of the part of the region above the axis, and B the area of the part below, $\int_a^b f(x)\, dx$ is defined to be $A - B$. In the Lebesgue definition of integration, we are allowed to use overestimates of these two areas, which are obtained by covering the corresponding parts of the region with an infinity of rectangles.

The letter $\mathscr{L}$ in the symbol $\mathscr{L}_2[a, b]$ is there because it is Lebesgue's initial. We are now in a position to state exactly what functions are in that space. To qualify, a function f must satisfy the following two conditions; (i) $\int_a^b [f(x)]^2\, dx$ is meaningful with the Lebesgue definition of integration, (ii) the value of this integral must be finite.

The first condition imposes no restriction in practice. For functions met in actual problems (as against theoretical constructions involving the axiom of choice), the Lebesgue integral is undefined only when we have an infinite area above the axis and also an infinite area below it. Then the definition leads to $A - B = \infty - \infty$, which is meaningless. As $[f(x)]^2$ is never negative, this difficulty cannot arise in connection with condition (i). The second condition does impose a real restriction. For instance, $1/\sqrt{x}$ gives a function that is excluded from $\mathscr{L}_2[0, 1]$, since

$$\int_0^1 (1/\sqrt{x})^2\, dx = +\infty.$$

If you are not familiar with the theory of Lebesgue integration, you may feel insecure and in the presence of a mystery when, as you surely will, you meet references to $\mathscr{L}_2$ spaces in the literature. This feeling is understandable, but it is not justified by the facts. Lebesgue's definition allows us to define integration for a wider class of functions than Riemann's does. When a function is integrable in the sense of Riemann, the Lebesgue definition agrees with Riemann's, so the familiar formulas of integration are preserved. An interesting discussion on how to use Lebesgue integration without knowing anything about it will be found in chapter 6 of Hardy, Littlewood and Polya.

A very helpful way of looking at $\mathcal{L}_2$ space is given by the Fourier coefficients. As was mentioned earlier, if $q_r(x) = \sqrt{(2/\pi)}\sin rx$, then $q_1, q_2, q_3, \ldots$ are mutually perpendicular unit vectors. They can be thought of as the basis of a co-ordinate system, and the Fourier coefficients, $f_r = (f, q_r)$, of any function f in $\mathcal{L}_2$, can be thought of as the co-ordinates of f in this system. It can then be shown that

$$\|f\|^2 = \sum_1^\infty f_r^2,$$

and for any two functions f, g in the space

$$(f, g) = \sum_1^\infty f_r g_r$$

Accordingly the formulas for length, distance, and scalar product closely resemble those in $\mathcal{E}^n$, but we are now summing from 1 to ∞ instead of from 1 to n.

A famous theorem, the Riesz–Fischer theorem of 1907, asserts that if we have any sequence of numbers, $f_1, f_2, f_3, \ldots$ for which $\sum_1^\infty f_r^2$ is finite, then there will be a function f in $\mathcal{L}_2[0, \pi]$ that will have this sequence as its Fourier coefficients. This result would not hold if we only accepted functions integrable in the sense of Riemann. For example, the function ϕ, that was described in connection with the repeated chopping up of a rectangle, has such a sequence of Fourier coefficients. With Riemann-integrable functions we would not be able to reproduce this sequence.

In chapter 2 we used the symbol ℓ_2 in connection with the metric of finite dimensional Euclidean space, and ℓ_2 thus can be used as a symbol for the space $\mathcal{E}^n$. The same symbol is used for the infinite-dimensional space, formed by unending sequences, $(x_1, x_2, x_3, \ldots)$ with $\|x\|^2 = \sum_1^\infty x_r^2$, this sum being required to be finite. This in fact is the commonest meaning attached to the symbol ℓ_2.

The Riesz–Fischer theorem says in effect that ℓ_2 and $\mathcal{L}_2$ are two different ways of describing the same space, which is also known as Hilbert space. There is some variation in the use of the term 'Hilbert space'. Some writers include finite dimensional Euclidean spaces under this heading, and some admit certain other infinite dimensional spaces. For all of them, the topic of central interest in

Hilbert space theory is the space we have been discussing. Hilbert spaces involving complex numbers are of great theoretical importance, but will not be touched on in this book.

The ℓ_2 description of Hilbert space looks so much like the description of $\mathscr{E}^n$ that there is a temptation to think of them as being practically the same thing. There are indeed helpful analogies between them, but caution is needed. The two next paragraphs illustrate this point.

In $\mathscr{E}^n$, a subspace is a different kind of thing from the whole space. In $\mathscr{E}^3$, for instance, a subspace may be a plane, a line, or a point—in each case, something of lower dimension than the whole space. In Hilbert space this is not so. If we impose the condition $x_1 = 0$ on the vectors $(x_1, x_2, x_3, \ldots)$ we define a subspace, but this subspace is a perfect replica of the whole space. If $(a_1, a_2, a_3, \ldots)$ is any vector in the whole space, there is a corresponding vector $(0, a_1, a_2, a_3, \ldots)$ in the subspace. The correspondence preserves lengths, distances, scalar products; it is like a rigid rotation, it establishes a congruence of the subspace with the whole space. Such a thing is totally impossible in $\mathscr{E}^n$.

Again, in space of three dimensions, the equation $x_1 + x_2 + x_3 = 1$ determines a plane. If a point is not in this plane, it is at a definite distance from it. It cannot be approached by any sequence of points in the plane. Now in the infinite dimensional space, ℓ_2, consider the points for which

$$\sum_1^\infty x_r = 1.$$

This certainly requires the series $\sum_1^\infty x_r$ to converge. Let us go even further, and require each vector considered to have some stage after which all the quantities x_r are 0. The origin does not satisfy this equation. For some fixed n, what point, $(x_1, x_2, \ldots x_n, 0, 0, 0, \ldots)$, that satisfies the equation, is nearest to the origin? It turns out to be $(1/n, 1/n, \ldots 1/n, 0, 0, 0, \ldots)$. The distance of this point from the origin is $1/\sqrt{n}$. By choosing n sufficiently large, we can make this as small as we like. Thus the origin, although it does not satisfy the equation, is the limit of points that do. In this respect, the origin is no better and no worse than any other point of the space, that does not satisfy the equation. Every point in the space is the limit of points that satisfy this equation. Thus we start by picturing the equation as defining something rather like a plane, and we end by not knowing what to think, for we seem to be dealing with something rather like a mist filling the entire space.

This difficulty does not arise with equations of the form $(x, c) = k$, where c is some fixed vector in ℓ_2. We cannot express $\sum_1^\infty x_r = 1$ in this way. For to do so, we would have to take $c = (1, 1, 1, 1, \ldots)$, and this is impossible, since $1^2 + 1^2 + 1^2 + 1^2 + \ldots = +\infty$, and so $(1, 1, 1, 1, \ldots)$ is not a vector in ℓ_2.

8.6. Applications of Hilbert space

Besides the uses of Hilbert space already mentioned, there are a number of other applications of interest to numerical analysts. These include approaches to problems known as the Rayleigh–Ritz method, Galerkin method, method of moments, variational methods.

The initial idea goes back to Rayleigh's work in 1870, long before Hilbert space had been thought of. Instead of looking for just any function, f, as the solution of a problem, Rayleigh restricted his attention to functions of the type $\sum_1^n c_r f_r$, where $f_1, \ldots f_n$ were specified functions, and the constants $c_1, \ldots c_n$ were to be chosen to make a certain expression a minimum. So, for instance, if the functions f_r had been powers of x, Rayleigh would have looked for the polynomial of a specified degree that came nearest to being a solution of his problem

The Hilbert space feature was added later. Suppose we are trying to solve $Tf = 0$. We try $f = \sum_1^n c_r f_r$. When we substitute this into the equation, we get $Tf = e$, where e represents the error in our output. It will of course depend on the constants c_r and on the variables that occur in the problem itself. We now want to put n conditions on e, to make it in some sense as small as possible, and to give us n equations that we can solve to find $c_1, \ldots c_n$. In the Galerkin method, we make e perpendicular to the subspace generated by $f_1, \ldots f_n$. That is, we make its projection on this subspace zero. The conditions for this are $(e, f_1) = 0 \ldots (e, f_n) = 0$. The problem thus reduces to solving these n equations. Incidentally, it would not be wise to do what was mentioned earlier, and choose $f_r(x) \equiv x^{r-1}$. This would lead to ill-conditioned equations. (See chapter 7, by C. T. H. Baker, in Delves and Walsh.)

We are not forced to use the subspace corresponding to the functions, f_r, that we mix to get f. We could get n equations by requiring the projection on some other subspace to be 0. So, if it serves our purpose, we may choose some sequence of functions, $g_1, g_2, \ldots$ and impose the conditions, $(e, g_1) = 0, \ldots (e, g_n) = 0$, from which the constants, c_r, can be deduced and an approximation to the solution found. Whether it is a good or a bad approximation clearly will depend on the wisdom with which the functions, f_r and g_r, have been chosen, both in their suitability to the solution itself, and in regard to the ill-conditioning or otherwise of the computations to which they lead. Here, as elsewhere, it is often helpful to make use of orthogonal systems of functions.

A very readable introduction to variational methods will be found in the book by S. H. Gould. Variational methods at a certain stage of their development call for a number of new concepts; it is therefore wise not to go too far in studying them until a certain familiarity with the basic ideas of Banach and Hilbert spaces has been achieved. Variational Methods are discussed in chapter 3 of Mikhlin and Smolitsky, *Approximate Methods for the Solution of Differential and Integral Equations*, (Elsevier, 1967). An idea of the scope of the applications of the variational method can be gained from *The Mathematics of Finite Elements*, edited by J. R. Whiteman (Academic Press, 1973).

9 Some tools of the trade

9.1. Geometry and generalization

INEQUALITIES pervade functional analysis, numerical analysis, and indeed analysis generally. If we wish to prove that k is the limit of a sequence, $a_1, a_2, a_3, \ldots$, it is rarely that we can put the exact value of $k - a_n$ in a form where it evidently tends to 0. Nearly always, we show $|k - a_n| \leqslant b_n$ where b_n clearly tends to 0. In this chapter, we consider ways of obtaining and proving inequalities. This may seem unnecessary, since most of the inequalities are standard ones, and the proofs can be found in books. Even so, it seems worth while. It is very much easier to remember an inequality if its meaning has been seen, and that meaning is something you would expect to be true on intuitive grounds. As to proofs, one of the best ways of fixing theorems in mind is continually to derive them, if possible by proofs in which a natural sequence of ideas is followed. This reduces reliance on pure memorizing and increases confidence and insight.

As was just mentioned, the meaning of an inequality is often enough to convince us of its truth. For instance, in $\mathscr{E}^3$, if θ is the angle between two vectors, u and v, we know $u \cdot v = \|u\| \|v\| \cos \theta$. As $\cos \theta$ never has magnitude exceeding 1, we are prepared to bet that

$$|u \cdot v| \leqslant \|u\| \|v\| \tag{1}$$

A formal proof of this will be helpful, particularly as we proceed towards generalizations. We can begin with this inequality as it appears in $\mathscr{E}^2$, where it takes the form

$$(u_1^2 + u_2^2)(v_1^2 + v_2^2) - (u_1 v_1 + u_2 v_2)^2 \geqslant 0. \tag{2}$$

We expect this to hold for all real values of the variables. If a polynomial can be expressed as a sum of squares, it can never take negative values. This gives us a hint for proving (2). Multiplying the left-hand side out, we find it comes to $(u_1 v_2 - u_2 v_1)^2$, which immediately proves the truth of (2).

We get a longer expression when we come to $\mathscr{E}^3$ for

$$(u_1^2 + u_2^2 + u_3^2)(v_1^2 + v_2^2 + v_3^2) - (u_1 v_1 + u_2 v_2 + u_3 v_3)^2$$
$$= (u_2 v_3 - u_3 v_2)^2 + (u_3 v_1 - u_1 v_3)^2 + (u_1 v_2 - u_2 v_1)^2. \tag{3}$$

This proves the inequality, and also suggests that we are going to get longer and longer expressions as we go to higher dimensions. We can check this for $\mathscr{E}^4$; we get six squares, corresponding to the six ways of choosing a pair from the four numbers, 1, 2, 3, 4. In fact the pattern for $\mathscr{E}^n$ gradually emerges. It can be expressed by writing

$$\left(\sum_{r=1}^{n} u_r^2\right)\left(\sum_{s=1}^{n} v_s^2\right) - \left(\sum_{r=1}^{n} u_r v_r\right)^2 = (1/2) \sum_{r=1}^{n} \sum_{s=1}^{n} (u_r v_s - u_s v_r)^2. \qquad (4)$$

On the right-hand side, the factor 1/2 is needed, as the summation causes each square to appear twice. This identity is most easily proved by starting with the right-hand side, and multiplying it out. The square terms give the first part of the left-hand side. The product

$$\sum_{r=1}^{n} \sum_{s=1}^{n} u_r v_s u_s v_r = \sum_{r=1}^{n} \sum_{s=1}^{n} u_r v_r u_s v_s = \left(\sum_{r=1}^{n} u_r v_r\right)\left(\sum_{s=1}^{n} u_s v_s\right) = \left(\sum_{r=1}^{n} u_r v_r\right)^2.$$

In $\mathscr{E}^3$, we expect equality to occur only when the two vectors have the same direction, or are in exactly opposite directions. Then $|\cos \theta| = 1$. Equation (3) confirms this algebraically for $\mathscr{E}^3$, and equation (4) shows that it is true for $\mathscr{E}^n$, since it is only when the vectors are proportional to each other that all the squares on the right-hand side of (4) will be 0.

In the introduction to Hardy, Littlewood and Polya, the authors offer some advice 'to readers who are anxious to avoid unnecessary immersion in detail'. Such readers (among other things) 'may take it for granted that "what goes for sums goes, with the obvious modifications, for integrals" (or vice versa)'.

If we apply this doctrine to the left-hand side of equation (4), replacing summation by integral signs, we are led to expect the inequality

$$\int_a^b [f(x)]^2 \, dx \cdot \int_a^b [g(x)]^2 \, dx - \left[\int_a^b f(x)g(x) \, dx\right]^2 \geq 0. \qquad (5)$$

This inequality, with f and g continuous functions, is in fact well known. It was published by H. A. Schwartz in 1885, and in many books is referred to as the Schwartz inequality. However this inequality had been published in the *Memoirs of the St. Petersburg Academy* in 1859, by V. Buniakowsky, and is correctly referred to as Buniakowsky's inequality. In view of this historical confusion, some authors use both names.

The integral analogy of the right-hand side of equation (4) is

$$(1/2) \int_a^b \int_a^b [u(x)v(y) - u(y)v(x)]^2 \, dx \, dy. \tag{6}$$

We can prove inequality (5), by showing that the expression on the left-hand side of (5) equals the expression (6) above, which clearly can never be negative.

This inequality holds for functions in $\mathcal{L}_2[a, b]$ as well as for continuous functions. In that space it gives the analogy of inequality (1), namely $|(u, v)| \leqslant \|u\| \|v\|$.

It is possible to stumble on a proof of an inequality in the course of working on some other topic. If a certain object is such that it can never be negative, and we have found an expression for that object, this automatically presents us both with an inequality for that expression, and with a way of proving it.

For instance, question (1) at the end of section 8.3 asks for the minimum value of $\|u - tv\|^2$, where u and v are given vectors. The answer is $[(u, u)(v, v) - (u, v)^2]/(v, v)$. The denominator here is positive. Since $\|u - tv\|^2$ can never be negative, the numerator can never be negative. But the numerator is a form of the expressions we have been considering in this section. In many books a short and ingenious proof of inequality (1) and its generalizations is given by this type of approach. It is quite possible that this proof was first thought of by someone who had calculated the minimum value of $\|u - tv\|^2$ and then saw that this gave a way of proving the inequality.

Bessel's inequality is another example of an inequality that is very plausible on geometrical grounds. As we saw earlier, the functions q_r, with $q_r(x) = \sqrt{(2/\pi)} \cdot \sin rx$, form an orthonormal set in $\mathcal{L}_2[0, \pi]$. If f is any function in this space, the projection of f onto the subspace generated by $q_1, \ldots q_n$ is

$$u = \sum_1^n f_r q_r, \quad \text{where} \quad f_r = (f, q_r),$$

the rth Fourier coefficient of f. As u is a projection of f, we expect u to be shorter than, or possibly just as long as, f. As the vectors $q_1, \ldots q_n$ are perpendicular and of unit length, the length of u is given by

$$\|u\|^2 = \sum_1^n f_r^2.$$

So we expect to have $\sum_1^n f_r^2 \leqslant \|f\|^2$. This is Bessel's inequality.

Geometry also indicates how we can obtain a more formal proof of this result. The projection u is obtained by dropping a perpendicular from f to the subspace. So, if $f = u + v$, the vector v will be perpendicular to the subspace, and in particular perpendicular to u. Hence, by Pythagoras' theorem, $\|f\|^2 = \|u\|^2 + \|v\|^2$. If we can compute $\|v\|^2$, that will give us the amount by which $\|f\|^2$ exceeds $\|u\|^2$, and being incapable of negative values, will prove the truth of the inequality. We proceed to do this. We have

$$\|v\|^2 = \left\| f - \sum_1^n f_r q_r \right\|^2 = \left(f - \sum_1^n f_r q_r, f - \sum_1^n f_s q_s \right)$$

$$= (f, f) - \left(\sum_1^n f_r q_r, f \right) - \left(f, \sum_1^n f_s q_s \right) + \left(\sum_1^n f_r q_r, \sum_1^n f_s q_s \right).$$

As $(q_r, q_s) = 0$ when $r \neq s$, and $= 1$ when $r = s$, the last bracket boils down simply to $\sum f_r^2$. In the third bracket, $(f, f_s q_s) = f_s(f, q_s) = f_s^2$, since (f, q_s) is the Fourier coefficient f_s. Thus the third bracket is $\sum f_s^2$. The same argument shows the second bracket to be $\sum f_r^2$. Finally, the first bracket, (f, f), is $\|f\|^2$. The right-hand side of the equation thus simplifies to $\|f\|^2 - \sum_1^n f_s^2$, which is what we expected. As $\|v\|^2$ cannot be negative, Bessel's inequality is proved.

As was mentioned in section 8.4, there are plenty of orthonormal systems every bit as good as the system $q_1, q_2, \ldots$. Bessel's inequality can be formulated so as to apply to any orthonormal system, as also can the Riesz–Fischer theorem, mentioned in section 8.5.

9.2. Calculus and inequalities

Calculus provides a useful method both for discovering and for proving inequalities. If we want to prove $f(x, y, z) \leqslant g(x, y, z)$, we can investigate the maximum value of $f(x, y, z) - g(x, y, z)$. If this maximum proves to be 0 or a negative quantity, the inequality is proved. Often it is more convenient to investigate the maximum value of $f(x, y, z)$ subject to the condition $g(x, y, z) = k$. The inequality is then proved, if the maximum is found to be k or some smaller quantity.

Finding a maximum or minimum, subject to a condition, can be reduced to the more usual type of calculus problem. For example, suppose we wish to find the point of the plane $x + 2y + 3z = 4$ in $\mathscr{C}^3$ that is nearest to the origin, by a calculus method. We want to minimize $x^2 + y^2 + z^2$. From the equation of the plane we can find x in terms of the other variables and substitute. We then have to minimize $\phi(y, z) = (4 - 2y - 3z)^2 + y^2 + z^2$, where y and z are not subject to any restriction, and this can be done by solving the equations $\partial\phi/\partial y = 0$, $\partial\phi/\partial z = 0$.

This procedure is elementary, but highly unsymmetrical and inelegant. Lagrange devised a method, known as the method of Lagrange multipliers, for solving such problems in a much more symmetrical and satisfying manner.

In the problem just considered, suppose the point (x, y, z) moves about in the plane with velocity (u, v, w). As $x + 2y + 3z$ is to remain constant, $0 = (d/dx)(x + 2y + 3z) = u + 2v + 3w$ is a condition imposed on the velocity. If $s = x^2 + y^2 + z^2$, $ds/dt = 2xu + 2yv + 2zw$. At a maximum or minimum, $ds/dt = 0$ for every permitted velocity. This means that, at the point (x, y, z) we are seeking, $2xu + 2yv + 2zw = 0$ whenever $u + 2v + 3w = 0$. This can happen only if the coefficients of u, v, w in the two equations are proportional, that is to say, if for all u, v, w, we have $2xu + 2yv + 2zw = \lambda(u + 2v + 3w)$, for some λ. This means $2x = \lambda$, $2y = 2\lambda$, $2z = 3\lambda$. As we have to have $x + 2y + 3z = 4$, this implies $\lambda = \frac{4}{7}$, from which the point (x, y, z) and its distance from the origin are easily found.

Lagrange's method can be extended to problems in which two or more conditions have to be satisfied.

This calculus approach is very general. It is always available, at least as a method of last resort, when other attacks have failed. It also has the merit that it leads to inequalities in a natural way. We do not have to know in advance what the inequality is.

For instance, consider a slight generalization of the problem just considered; minimize $x^2 + y^2 + z^2$ subject to $ax + by + cz = k$. The minimum is found to be $k^2/(a^2 + b^2 + c^2)$. This means that $x^2 + y^2 + z^2 \geq k^2/(a^2 + b^2 + c^2)$, which is equivalent to $(x^2 + y^2 + z^2) \times (a^2 + b^2 + c^2) \geq k^2$. Substituting for k, we have $(x^2 + y^2 + z^2) \times (a^2 + b^2 + c^2) \geq (ax + by + cz)^2$. We have arrived at the inequality for the scalar product, discussed in section 9.1, and usually known as

164 *Some tools of the trade*

Cauchy's inequality. This approach would work equally well in $\mathscr{E}^n$ and lead to the generalized inequality.

All such inequalities can be approached from two directions. Instead of minimizing $\sum x_r^2$, subject to $\sum a_r x_r = \text{constant}$, we could look for the extreme values of $\sum a_r x_r$, subject to $\sum x_r^2 = \text{constant}$. Here we are seeking the extreme values of (a, x) with a and $\|x\|$ prescribed. We would arrive at the maximum value $\|a\|\,\|x\|$ and the minimum value $-\|a\|\,\|x\|$, these arising, as we would expect, when x has the same direction as a, and when it is opposite to a.

Exercise. Verify this statement.

We could also arrive at the triangle inequality directly, without using any other inequality as a stepping stone, as happens in a frequently used proof.

We want to show $\|a + x\| \leqslant \|a\| + \|x\|$. We may look at this question in the following way Let $\|x\| = r$. Then $a + x$ is a point somewhere on the sphere $S(a, r)$. We want to show that no point of this sphere can be further from the origin than the distance $\|a\| + r$. So we pose the problem, what is the maximum value of $\|a + x\|$, if a is a given vector, and $\|x\| = r$ is a condition? So we consider the extreme values of

$$\sum_1^n (a_s + x_s)^2,$$

subject to

$$\sum_1^n x_s^2 = r^2.$$

This leads to the very simple equations $a_s + x_s = \lambda x_s$, and λ can be found by substituting for x_s in the imposed condition. Two values of λ result, for there is a point of the sphere nearest to the origin as well as one farthest away.

The Lagrange method tells us when the value is stationary. Without further elaboration, it does not tell us whether we have found a maximum, a minimum, or simply a hesitation. Usually, the geometry of the situation is such that it is clear which of these we have in fact located.

Hölder's inequality was considered in chapter 4. This inequality arises quite naturally in the study of functionals, $\ell_p \to \mathbb{R}$. Let the functional f map

$$(x_1, \ldots x_n) \to \sum_1^n a_r x_r.$$

As in chapter 4, it is sufficient if we restrict ourselves to the case where $a_r \geq 0$ and $x_r \geq 0$ for all r. In these circumstances $\|f\|$ is the maximum value of $\sum_1^n a_r x_r$ subject to $\|x\| = 1$, which means $\sum_1^n x_r^p = 1$.

Writing $u_r = dx_r/dt$, we want to make $\sum a_r u_r = 0$ for all $(u_1, \ldots u_n)$ that satisfy $\sum x_r^{p-1} u_r = 0$. This last equation comes from differentiating the condition on x, and removing a factor p. So we want $a_r = \lambda x_r^{p-1}$ for each r. Hence $x_r = (a/\lambda)^{1/(p-1)}$. Substituting this in the condition given, we have $1 = \sum x_r^p = \sum (a_r/\lambda)^{p/(p-1)}$. The mysterious number, $q = p/(p-1)$ has come in quite naturally. Using q as an abbreviation for $p/(p-1)$, the last equation may be written $1 = \sum (a_r/\lambda)^q$, from which it follows $\lambda^q = \sum a_r^q$.

The multiplier λ that arises in the Lagrange method is usually of some significance, and this case is no exception. If we multiply the equation $\lambda x_r^{p-1} = a_r$ by x_r and sum over r, we obtain $\lambda \sum x_r^p = \sum a_r x_r$. As $\sum x_r^p = 1$, this implies $\lambda = \sum a_r x_r$, so λ is no other than the maximum value that we are seeking. As $\lambda^q = \sum a_r^q$, we can write $\lambda = \|a\|_q$, the norm of the vector a in the space ℓ_q. This then is the desired value of f. The mapping f does not multiply the magnitude of any vector by more than $\|a\|_q$. As $f(x) = \sum a_r x_r$, we are entitled to conclude that $|\sum a_r x_r| \leq \|x\|_p \|a\|_q$, and this is the concise form of Hölder's inequality.

We can arrive at the same inequality by posing the question the other way round; what is the minimum distance from the origin, in the ℓ_p metric, of a point x that satisfies $\sum_1^n a_r x_r = c$? This would probably be the most natural way of arriving at Hölder's inequality, in connection with the proof of the triangle inequality for ℓ_p that was discussed in section 4.2.

It would also be possible to attack the triangle inequality in ℓ_p directly, in the way already outlined for the Euclidean case, by maximizing $\|a + x\|$ for given a and prescribed $\|x\|$. It is of interest that Minkowski himself did not use the algebraic proof discussed in chapter 4, but proved the triangle inequality for his spaces by a calculus approach.

9.3. Continuity; distance, norm, scalar product

In any metric space, the distance function is automatically continuous, in the sense that small changes in u and v produce small changes in $d(u, v)$. Intuitively, this is immediately evident. If u and v

FIG. 39

are displaced through small distances, r and R respectively, from their original positions, u_0 and v_0, their new positions will lie on the spheres $S(u_0, r)$ and $S(v_0, R)$. It is clear from Figure 39 that the distance between them cannot have increased by more than $r + R$, nor have decreased by more than this same amount. The formal proof, by means of the polygon inequality, is easy.

In a normed space, $\|v\| = d(0, v)$. The norm thus depends continuously on the vector involved. If $v_n \rightarrow v$, it follows that $\|v_n\| \rightarrow \|v\|$. A proof of this was promised in section 5.6.

Finally, in Hilbert space, the scalar product, (u, v), depends continuously on u and v. An adjustment of the values of (u, v) could be made in two stages; first alter u a little, then alter v a little. If the change in each of these stages is small, the overall change will be small. Suppose then, v changes to $v + h$. We have $(u, v + h) - (u, v) = (u, h)$ by the algebraic properties of the product. In section 9.1 we had the inequality $|(u, h)| \leq \|u\| \|h\|$. It is clear that $(u, h) \rightarrow 0$ if $h \rightarrow 0$. This proves what we want for small changes in v; exactly similar considerations apply to small changes in u, and the result is proved.

10 Some bridgeheads

10.1. Functionals and the dual space

THIS CHAPTER is in the nature of fragmentary notes, which indicate the existence of parts of mathematics without in any way claiming to explore them. The themes involve functionals, dual spaces, convexity and compactness. The final section is perhaps rather out of place under this heading, for it describes a road that leads nowhere.

A special case of the space $\mathscr{B}(\mathscr{X}, \mathscr{Y})$ occurs when the output space, $\mathscr{Y}$, is $\mathbb{R}$, the space of real numbers. This space is known as the dual space of $\mathscr{X}$, and is invariably denoted by the symbol $\mathscr{X}^*$. Its elements, as has been mentioned earlier, are called bounded, linear functionals. The usual operator norm, of course, applies to them, so that, if L is such a functional

$$\|L\| = \sup\{|Lx|; \|x\| = 1\}.$$

There are foreshadowings of the dual space in classical mathematics. A conic is usually regarded as a set of points, but it can equally well be seen in the dual sense, as the set of its tangents. The line $ax + by = c$ is a tangent to the circle $x^2 + y^2 = 1$ if $a^2 + b^2 = c^2$, and the latter equation has just as much right to be regarded as the equation of the circle as the former. In projective geometry, every theorem about points (x, y, z) has a companion theorem, its dual, about (a, b, c), a vector that specifies the line with homogeneous equation $ax + by + cz = 0$.

If the space $\mathscr{X}$ is finite dimensional, with points $(x_1, \ldots x_n)$, every L in $\mathscr{X}^*$ is specified by $Lx = \sum_1^n a_r x_r$, and may be identified by giving $(a_1, \ldots a_n)$, the vector formed by its coefficients. In Euclidean geometry the significance of the dual space is camouflaged by the fact that it is identical copy of the original space. As we saw in chapter 9, the maximum value of $\left|\sum_1^n a_r x_r\right|$, subject to $\sqrt{\left(\sum_1^n x_r^2\right)} = 1$, is $\sqrt{\left(\sum_1^n a_r^2\right)}$, and this gives $\|L\|$. Thus in $\mathscr{X}^*$, $\|L\|$ is given by a formula exactly corresponding to that for $\|x\|$ in $\mathscr{X}$.

A distinction appears when the dual space of ℓ_p is considered. Hölder's inequality, as was shown in section 9.2, can be put in the concise form $|\sum a_r x_r| \leq \|x\|_p \|a\|_q$, where $(1/p) + (1/q) = 1$. Equality can be attained. It follows that $\|L\| = \|a\|_q$. So in the dual space, the formula for the norm is that appropriate to ℓ_q. So ℓ_p^*, the dual of ℓ_p, is not the same as ℓ_p, but is a different space, ℓ_q. In view of the symmetry of the relation between p and q, $\ell_q^* = \ell_p$. Each space is the dual of the other. It should not be assumed that something of this kind will always happen; finding the dual space of $\mathscr{X}^*$ does not always bring us back to $\mathscr{X}$. In the discussion above, we were concerned with the finite dimensional space ℓ_p. It can be shown that the result, $\ell_p^* = \ell_q$, still holds when the space is infinite dimensional.

If w is a function continuous on $[0, 1]$, we can use it to define a bounded linear functional, $\mathscr{C}[0, 1] \to \mathbb{R}$, by making

$$f \to \int_0^1 f(t)w(t)\, dt.$$

Here w plays the role of a weight function, attaching different amounts of importance to different parts of the interval $[0, 1]$. However it is impossible to choose a function w that concentrates all the weight at one point, say $t = c$, and makes the output equal to $f(c)$. However, there is a perfectly good functional that does this, $L: f \to f(c)$. Thus certain things can be done with the help of the dual space that are impossible with the original space.

Luenberger gives an example of this in an optimization problem. How should the rate of burning be arranged if a rocket is to be fired vertically to a prescribed height, in a vacuum, with the minimum expenditure of fuel? It turns out that the burning of fuel should be concentrated as near the start of the motion as possible; the theoretical answer is that there should be an infinite rate of burning for zero time. This theoretical solution, which again involves the question of something being concentrated at a point, can be reached only if the problem is posed in the dual space (see Luenberger, p. 5 and pp. 125–6.) Luenberger in fact speaks of 'the general rule ... that minimum norm problems must be formulated in a dual space if one is to guarantee the existence of a solution'.

You may occasionally come across references to Dirac's delta function. Although they knew quite well that there was no weight function, w, that sent the input f to the output $f(c)$, physicists decided to work formally with a function $w(t) = \delta(t - c)$, that was assumed to do just that. On this rather uncertain basis, they obtained certain results. Later, the logical difficulties were cleared up by the theory of distributions, in which this and other apparently paradoxical concepts were legitimized by considering them as functionals in the dual space rather than as functions in the original space. The theory of distributions appears to have considerable possibilities. It is not hard to understand once the concept of functional has been thoroughly digested. A clear and interesting account of it has been given in a little pamphlet, *Introduction to the Theory of Distributions*, by I. Halperin (University of Toronto Press, 1952).

Dual spaces provide a most useful device for dealing with maximum and minimum problems. Many significant problems can be stated in terms of finding the minimum distance from the origin to some convex region in a Banach space. Figure 40 shows the situation as it would appear in a space of only two dimensions. We want to find $\|OP\|$, where P is the point of the shaded oval nearest to O. We want to pin this value down between two numbers, an overestimate and an underestimate. We can get an overestimate by choosing a suitable point Q in the shaded region, for we know $\|OP\| \le \|OQ\|$. To get an underestimate, we consider any line that has O on one side of it, and the shaded region on the other side. The line shown dotted has this property. If M is the point of this line nearest to the origin, it will be seen that $\|OM\| \le \|OP\|$, so $\|OM\|$ provides an underestimate. This example gives a very faint indication of the general principle that a maximum problem in $\mathscr{X}^*$ corresponds to a minimum problem in $\mathscr{X}$, and by taking approximations in the two spaces we are able to approach the desired value both from above and below, so that we know the size of our error.

A line such as the dotted line in Figure 40, which contains at least one boundary point of a convex region and has the convex region entirely on one side, is called a *support* of the convex region. In three dimensions a support of a convex region is a plane with the

Fig. 40

corresponding property, and in higher dimensions an analogous object, a hyperplane with such a property.

As an aid to visualizing the meaning of a functional, it is again helpful to start with the situation in two dimensions. Suppose then the space $\mathscr{X}$ is a plane. Then every functional is of the form $L:(x, y) \to ax + by$. Every point of the plane thus has a number assigned to it by L, as would happen, for instance, if every point had its own temperature. The isotherms could be drawn, connecting points with the same temperature, that is to say, with the same value of $L(x, y)$. Points having the value c would lie on the line $ax + by = c$. Thus we would see a system of parallel lines. If (x, y) lies on the line, $ax + by = 1$, (cx, cy) will lie on the line, $ax + by = c$, so the whole system is determined and L is fixed as soon as the line $ax + by = 1$ has been drawn. $\|L\|$ can be defined in terms of this line. Instead of seeking the maximum of $L(x, y)$ on the unit sphere, $S(0, 1)$, we seek the smallest value of r for which $S(0, r)$ meets the line $ax + by = 1$. Then $\|L\| = 1/r$. Thus, if $\|L\| = 1$, this means that the line $ax + by = 1$ does not enter the interior of $\bar{B}(0, 1)$ but that it does contain a boundary point of that ball. In other words, this line must be a support of the convex region, $\bar{B}(0, 1)$. This condition generalizes, $\|L\| = 1$ means that the equation $Lv = 1$ gives a support of $\bar{B}(0, 1)$. This result will be used in the following section.

10.2. An existence theorem

Theorem 2 of section 6.3 was stated without proof. It said that, if a was any vector in a Banach space, $\mathscr{S}$, there was a linear functional, $L:\mathscr{S} \to \mathbb{R}$, with $La = \|a\|$, and $\|L\| = 1$.

We follow our usual practice, and consider how this would look in two dimensions. Let $a = (1, 0)$. The space $\mathscr{S}$ may be the plane with metric of ℓ_1, ℓ_2 or ℓ_∞. In each of these $\|a\| = 1$, so we require $La = 1$. This means that $L(x, y) = x + ky$ for some k. To make $\|L\| = 1$, we have to ensure that $x + ky = 1$ is the equation of a support line of $\bar{B}(0, 1)$. This line automatically goes through $(1, 0)$.

For the space ℓ_2, the unit ball is bounded by the circle, $x_2 + y_2 = 1$. The only line through $(1, 0)$ that does not go inside this circle is the tangent, $x = 1$. So we must choose $k = 0$, and L is $(x, y) \to x$.

The situation for ℓ_∞ is similar. $S(0, 1)$ is then a square, with sides in the lines $x = \pm 1$, $y = \pm 1$. The only line through $(1, 0)$ that is a support line for the unit ball is again $x = 1$.

For ℓ_1, the sphere $S(0, 1)$, is a tilted square. The sides that pass through $(1, 0)$ lie in the lines $x + y = 1$ and $x - y = 1$. The line

$x + ky = 1$ will avoid the interior of this square if $-1 \leq k \leq 1$. Thus in this case there are many ways of choosing L to satisfy the conditions. Notice that Theorem 2, and the use made of it in section 6.3, did not depend on there being only one such L. It was sufficient if at least one L with these properties existed.

These three cases had something in common. In each case $\bar{B}(0, 1)$ was convex and we sought a support line for it. We saw in section 4.3 that in any normed vector space, $\bar{B}(0, 1)$ must be convex. There is a general theorem that, through any point on the boundary of a convex region, a support can be drawn. Our experience of two and three dimensions makes this extremely plausible—that makes it easy to visualize and remember—and it can in fact be proved. The equation of such a support will be of the form $Lv = \text{constant}$, where L is a bounded, linear functional.

Now for any vector a, $\neq 0$, the point, $a/\|a\|$, lies on $S(0, 1)$, the boundary of $\bar{B}(0, 1)$. Let $Lv = 1$ be the equation of the support for $\bar{B}(0, 1)$ that passes through $a/\|a\|$ (or a support, if there are many). By the remark at the end of section 10.1, $\|L\| = 1$. Since $a/\|a\|$ satisfies $Lv = 1$, it follows that $1 = L(a/\|a\|) = (1/\|a\|)La$, so $La = \|a\|$ as required.

This result can be seen in terms of extending the domain of definition of a functional. If we confine our attention to points on the line joining the origin to a, the functional $xa \rightarrow x \|a\|$ is clearly linear, has norm 1, and maps $a \rightarrow \|a\|$. Theorem 2 in effect asserts that, without changing the mapping on this line, we can extend the definition of this functional to the whole space and keep these properties. This theorem indeed is usually proved as a corollary to a more general and very famous theorem, the Hahn–Banach theorem, which is concerned with extending the definition of a functional from part of a space to the whole space. The Hahn–Banach theorem depends on the same kind of ideas as those just discussed—the existence of supports for convex regions.

Luenberger calls the Hahn–Banach theorem 'the most important theorem for the study of optimization in linear spaces', (see Luenberger, p. 110). His book is clearly written and has the great merit of emphasizing and bringing out the geometrical meaning of the processes and results that are covered.

10.3. An application of functionals

An intriguing formula is given in section 4.5 of Dahlquist and Björck. This formula is extremely simple to prove and yet has varied and interesting uses. It concerns a bounded, linear functional $L : \mathscr{C}[a, b] \to \mathbb{R}$. It is assumed that $Lp = 0$, if p is any polynomial function of degree N or less. We then have the theorem; for any function, f, continuous on $[a, b]$ and for any polynomial of degree not exceeding N, $|Lf| \leqslant \|L\| \, \|f - p\|$.

The proof is immediate. $Lp = 0$, so $Lf = Lf - Lp = L(f - p)$ since L is linear. So $|Lf| = |L(f - p)| \leqslant \|L\| \, \|f - p\|$.

As an application of this, consider L defined by $Lf = f(0) - 3f(1) + 3f(2) - f(3)$, where f is continuous on the interval $[0, 3]$. Clearly L is linear. It is bounded, for if $\|f\| = 1$, this means $-1 \leqslant f(x) \leqslant +1$, and the largest value we can get for Lf comes by taking $f(0) = 1$, $f(1) = -1$, $f(2) = 1$, and $f(3) = -1$, which makes $Lf = 8$. Hence $\|L\| = 8$. If p is any quadratic function, $Lp = 0$, so we may take $N = 2$. The theorem tells us $\|f - p\| \geqslant |Lf|/8$. That is, it tells us there is a certain, unavoidable error we are bound to make if we try to fit the function f by a quadratic.

The table below gives some data for a function, f, which is clearly highly unsuitable for fitting by a quadratic. We expect the formula to predict appreciable, unavoidable errors.

x	0	1	2	3
$f(x)$	1	-2	4	-8

Here $Lf = f(0) - 3f(1) + 3f(2) - f(3) = 1 + 6 + 12 + 8 = 27$. So for any quadratic, p, we shall have $\|f - p\| \geqslant \frac{27}{8} = 3.375$. In fact taking $p(x) \equiv -2.25x^2 + 6x - 2.375$ gives the following errors:

x	0	1	2	3
$e(x)$	3.375	-3.375	3.375	-3.375

The practical conclusion, of course, is what we expected—the errors are too large for the approximation to be worth the trouble taken to find it. A theoretical conclusion also can be drawn. The quadratic used achieves the predicted unavoidable error, so no other quadratic can do better. In other circumstances, with errors of a reasonable size, it might be useful to know that further search would be unprofitable, that we had already reached, or perhaps come very near to, the best possible quadratic. In this particular case, we can see that the quadratic cannot be improved on by using a theorem of Chebyshev about polynomials of best fit.

The same formula, $|Lf| \leq \|L\| \|f - p\|$, also has applications to numerical integration. Dahlquist and Björck give a simple example of this. The trapezium rule approximates to $\int_0^h f(x) \, dx$, the area under a curve, by finding the area under a chord of that curve, namely $(h/2)[f(0) + f(h)]$. Let Lf denote the error made by so doing, so

$$Lf = (h/2)[f(0) + f(h)] - \int_0^h f(x) \, dx.$$

The functional L is linear. Also $Lp = 0$ if $p(x) \equiv mx + c$, for then the curve and the chord coincide. Hence we may apply the formula with p a polynomial of degree 1 or less. We find $\|L\| = 2h$. The value of $\|f - p\|$ is estimated from information about the magnitude of the second derivative, f''. It is then shown that the error, $|Lf|$, cannot exceed a certain amount. Thus, in this case the formula works in the opposite direction. It shows us the efficiency, rather than the limitations, of the method.

10.4. Compactness

The first paper published by Maurice Fréchet, when he set out to make an extensive generalization of real variable theory, introduced the concept of compactness. This paper, two pages in length, was published in the *Comptes Rendus* for 1904. Two strands of thought led to this idea, one coming from inside real variable theory, the other from outside.

The second strand involved variational methods of importance both in mathematical theory and in computational practice. Many problems in physics are equivalent to making some quantity a maximum or a minimum. A statical system settles down so as to make potential energy a minimum; a soap bubble on a wire frame forms a surface of minimum area; electric currents flow through a network in the way that generates least heat. The mathematical problems corresponding to, or suggested by, these situations are thus equivalent to finding the appropriate maximum or minimum. Very beautiful theories were based on this approach. However, Weierstrass pointed out a serious difficulty. It is no good looking for a minimum, if a minimum does not exist. Even if you are able to prove $Tf \geq m$ and get values of Tf as near to m as you like, it does not follow that there will be some f_0 that makes $Tf_0 = m$. This is expressed by saying that the minimum may not be attained. For example, let

$$Tf = \int_{-1}^{1} x^2 [f'(x)]^2 \, dx.$$

Here f is to be continuous and have a continuous derivative on the interval $[-1, 1]$, and there are boundary conditions $f(-1) = -1, f(1) = 1$. Clearly $Tf \geqslant 0$. We can get values of Tf as small as we like by taking $f(x) = -1$ from $x = -1$ to a value of x just less than 0; then let $f(x)$ rise swiftly from -1 to $+1$, and make $f(x) = 1$ from a value of x just more than 0 until $x = 1$ is reached. Yet clearly, there is no f that makes $Tf = 0$, for that would require $f'(x) \equiv 0$. This would mean f constant, and it would be impossible to satisfy both the boundary conditions.

The strand coming from real variable theory was also connected with Weierstrass, who had a theorem that a continuous function on a finite interval, $[a, b]$, attains its maximum and minimum. This means that, if

$$m = \inf\{f(x) : a \leqslant x \leqslant b\},$$

there is some x_0 in the interval, such that $f(x_0) = m$. In the same way, there is some x_1 that makes $f(x_1) = M$, where M is the supremum. It can also be proved that M and m must be finite.

Neither of these things need be true for an infinite interval, or for an open interval, that is, one defined by $a < x < b$. If the interval consists of all real values and $f(x) \equiv \tan^{-1} x$, then $M = \pi/2$ and $m = -\pi/2$, but there is no number that gives these values for $\tan^{-1} x$. Again, if $f(x) \equiv x/(1 - x^2)$, then f is continuous for $-1 < x < 1$, but $M = +\infty$ and $m = -\infty$.

The type of argument used to prove the theorems for the finite, closed interval $[a, b]$ is as follows. Since m is the infimum, however large the whole number n may be, there must be x_n in the interval such that $m \leqslant f(x_n) \leqslant m + (1/n)$. We cannot conclude that the sequence $x_1, x_2, x_3, \ldots$ tends to a limit, because there may be many different places in the interval where the value comes close to m. Some x_n may be close to one such place, some to another, so the sequence as a whole may not converge. The Bolzano–Weierstrass theorem assures us that it is possible to cross out some of the terms in the sequence in such a way that the surviving terms converge to some x_0 in the interval. It is then shown that $f(x_0) = m$. This automatically proves that m is finite, for f is a function, $\mathbb{R} \to \mathbb{R}$, and $-\infty$ is not a member of $\mathbb{R}$.

This argument breaks down in the case of an infinite or an open interval.

Fréchet thus saw that the crucial point was whether a convergent subsequence could be selected from a sequence $x_1, x_2, x_3, \ldots$. If we knew this could be done with any sequence consisting of points in some part of a metric space, $\mathcal{X}$, then we would be able to prove theorems about continuous functions, $\mathcal{X} \to \mathbb{R}$, corresponding to the theorems just mentioned for continuous functions, $\mathbb{R} \to \mathbb{R}$. Cheney mentions that many problems in approximation are equivalent to finding the point in some part of a metric space nearest to some given point. This is a problem about minimum distance. It is obviously important to know whether the minimum is attained— whether there is such a thing as 'the nearest point'.

Fréchet called a region compact if it had this property, of allowing a subsequence, converging to a point in the region, to be selected from any sequence of points located in the region.

A set of points in a metric space is called bounded if it lies in any ball, $\bar{B}(O, r)$. In $\mathbb{R}$, or in any finite dimensional space, $\mathbb{R}^n$ or $\mathscr{C}^n$, any bounded set is automatically compact. If you put an infinity of points into, say, the ball $\bar{B}(0, 1)$ in two dimensions, there is bound to be at least one point with an infinity of these points in any circle with it as centre.

It is entirely different with spaces of infinite dimension. Consider, for instance, the sequence, $f_1, f_2, f_3, \ldots$ in $\mathscr{C}[0, \pi]$, with $f_n(x) \equiv \sin nx$. For large n, their graphs oscillate very fast. There is no way of selecting a sequence of them that converges to a continuous function. Yet for each n, $\|f_n\| = 1$, so all of them lie in $\bar{B}(0, 1)$. The trouble here is that $f_n'(x) = n \cos nx$, so that, by taking n sufficiently large, we can get graphs that are arbitrarily steep. Work published in the years 1883–1895 by the mathematicians, G. Ascoli and C. Arzelà, showed that compactness could be assured by requiring a property called equicontinuity in the family of functions. This has the effect of limiting the steepness of the permitted graphs. If the functions, f_n, have derivatives, f_n', and if for some constant, c, the inequality $|f_n'(x)| \le c$ holds for all n and for all x in the interval, this is sufficient to guarantee equicontinuity. It is not necessary. Functions can be equicontinuous without being differentiable.

The theory of integral equations is helped by the fact that, if the kernel K is continuous, and if f lies in some ball in $\mathscr{C}[0, 1]$, then the region consisting of all the functions, g, given by

$$g(x) = \int_0^1 K(x, y)f(y)\, dy,$$

is compact.

In Hilbert space the unit ball, $\bar{B}(0, 1)$, is not compact. Let v_n denote the vector, $(0, 0, \ldots, 0, 1, 0, 0, \ldots)$ in ℓ_2, where all the entries are 0, except the nth, which is 1. It is impossible to select a convergent subsequence from the sequence whose nth term is v_n. For if m and n are any two distinct whole numbers, $\|v_m - v_n\| = \sqrt{2}$. However long we wait, it is impossible to find v_n, such that the tail of the subsequence following it, is contained in a ball of radius less than $\sqrt{2}$. Thus it is impossible to find a subsequence that satisfies the Cauchy condition.

A consequence of this is that in Hilbert space, the theory of linear transformations, $x \to y$, with

$$y_r = \sum_{s=1}^{\infty} a_{rs}x_s,$$

and the corresponding theory of quadratic forms,

$$\sum_{r=1}^{\infty} \sum_{s=1}^{\infty} a_{rs}x_rx_s,$$

differs in many ways from the theories for finite-dimensional spaces.

10.5. Vector products

We have seen that the two basic operations of vector theory, giving $u + v$ and ku, have a very wide range of applications. When the scalar product, $u \cdot v$ or (u, v) is included, the applications, though less wide, are still extensive. Our thoughts naturally turn to the remaining operation of elementary vector theory, the vector product. What scope has it for generalization?

The answer appears to be none. It would be very rash to say that any mathematical concept is totally incapable of leading to generalization; there is the question of how much alteration is to be permitted. Vector products can be generalized by bringing in anti-symmetric tensors, or by defining a vector in n dimensions in terms of $(n-1)$ given vectors. All that is asserted now is that the obvious, direct generalization of vector product is impossible. In more than three dimensions, there is no relationship, $w = f(u, v)$, that has the properties of a product.

The algebraic properties of the vector product arise from the fact the equations

$$w_1 = u_2 v_3 - u_3 v_2, \quad w_2 = u_3 v_1 - u_1 v_3, \quad w_3 = u_1 v_2 - u_2 v_1,$$

which define $w = u \times v$, are linear in u and v separately; they define a bilinear function, $(u, v) \to w$. There is no difficulty at all in generalizing this. We have only to take

$$w_r = \sum_s \sum_t c_{rst} u_s v_t$$

with any constants c_{rst} we please to get such a bilinear function. The snag is that the resulting w does not depend only on the vectors u and v; it depends also on the co-ordinate system being used. The mapping is from two vectors, u, v *and* a co-ordinate system, to w.

In three dimensions, w is perpendicular to the plane of u and v, its magnitude is given by the area of the parallelogram with sides u and v, and the sense of w is determined by a right-handed screw, turning from u to v. As right-angles are involved, we naturally require the co-ordinate system to be orthonormal. As right-handedness is involved, we do not expect the definition to be unaltered by reflection. However, if w is given by the equations above in terms of u and v, and if a rotation about the origin changes these vectors to U, V, W respectively, then W will be given in terms of U and V by precisely the same equations. It is this property, combined with bilinearity, that does not generalize.

To prove this, let $e_1, e_2, \ldots e_n$ be an orthonormal system in $\mathscr{E}^n$ Let

$$e_1 \times e_2 = \sum_1^n k_r e_r.$$

We write this as

$$e_1 \times e_2 = k_1 e_1 + k_2 e_2 + \sum_3^n k_r e_r.$$

Now $e_1 \to -e_1$, $e_2 \to -e_2$, the other vectors remaining unchanged, defines a rotation in $\mathscr{E}^n$. Since rotations are to leave the equations unaffected, we may substitute the new values in the equation above. This gives

$$e_1 \times e_2 = -k_1 e_2 - k_2 e_2 + \sum_3^n k_r e_r.$$

From this it follows that $k_1 = k_2 = 0$, so we may write

$$e_1 \times e_2 = k_3 e_3 + \sum_4^n k_r e_r.$$

Now let the rotation act, with $e_2 \rightarrow -e_2$, $e_3 \rightarrow -e_3$, and the other e_r unchanged. This gives

$$-e_1 \times e_2 = -k_3 e_3 + \sum_4^n k_r e_r.$$

Comparing the last two equations, we see that $k_r = 0$ for $r \geq 4$, so $e_1 \times e_2 = k_3 e_3$. If we are in 4 dimensions or more, we can now consider the rotation with $e_2 \rightarrow -e_2$, $e_4 \rightarrow -e_4$, and the remaining e_r unchanged. This shows that $-e_1 \times e_2 = k_3 e_3$, and so $k_3 = 0$. Thus $e_1 \times e_2 = 0$. There is nothing special about e_1 and e_2. In the same way we show that the vector product of any two of the basis vectors is 0. The same kind of argument shows that $e_1 \times e_1 = 0$, and similarly for the other e_r. It follows that all vector products are 0. This is the only definition that meets the conditions, and it is clearly useless.

Bibliography

BEREZIN AND ZHIDKOV. *Computing methods* (Pergamon, 1965).

E. W. CHENEY. *Introduction to approximation theory* (McGraw Hill, 1966).

C. W. CLENSHAW.
1. *Chebyshev series for mathematical functions* (National Physical Laboratory. Mathematical tables, Volume 5. 1956).
2. A comparison of the 'best' polynomial approximations with the truncated Chebshev series expansions (*Society for Industrial and Applied Mathematics. Journal for numerical analysis*, series B, Volume 1, pp. 26–37. 1964).
3. Chapters 2 and 3 of the book by J. G. Hayes, which see below.

L. COLLATZ. *Functional analysis for numerical analysis* (Academic Press, 1966). Beginners may find difficulty with parts of this book, owing to metric spaces being approached through the more general idea of pseudometric spaces.

G. DAHLQUIST AND A. B. BJÖRCK. *Numerical methods* (Prentice Hall, 1974).

P. J. DAVIS. *Interpolation and approximation* (Blaisdell, 1963).

L. M. DELVES AND J. WALSH. *Numerical solution of integral equations* (Clarendon Press, Oxford, 1974).

L. FOX AND D. F. MAYERS. *Computing methods for scientists and engineers* (Clarendon Press, Oxford, 1968).

S. H. GOULD. *Variational methods for eigenvalue problems* (*Mathematical expositions*, No. 10. University of Toronto).

I. GRATTAN-GUINNESS. *The development of the foundations of mathematical analysis from Euler to Riemann* (M.I.T., 1970).

D. C. HANDSCOMB. *Methods of numerical approximation* (Pergamon, 1968).

HARDY, LITTLEWOOD AND POLYA. *Inequalities* (Cambridge University Press, 1934).

J. F. HART. *Computer approximations* (Wiley, 1968).

J. G. HAYES. *Numerical approximations to functions and data* (Athlone, 1967).

KANTOROVICH AND AKILOV. *Functional analysis in normed spaces* (Pergamon, 1964).

BINH LAM AND DAVID ELLIOTT. On a conjecture of C. W. Clenshaw (*Society for Industrial and Applied Mathematics. Journal*. Volume 9, p. 44. 1972).

D. G. LUENBERGER. *Optimization by vector space methods* (Wiley, 1969).

PATRICIA PRENTER. Polynomial operators and equations. This is a chapter in the book *Non-linear functional analysis and applications*, edited by L. B. Rall, (Academic Press, 1971).

L. B. RALL.
1. Quadratic equations in Banach space (*Rendiconti del Circolo Matematico di Palermo*, volume 10, pp. 314–332, 1961).

2. *Computational solutions of non-linear operator equations* (Wiley, 1969).

J. TODD.

1. The condition of finite segments of the Hilbert matrix (*National Bureau of Standards, U.S.A., Applied mathematics series*, No. 39. 1954).

2. *Survey of numerical analysis* (McGraw Hill, 1962).

Index